Disaster
Management

A guide to issues management and crisis communication

CHRIS SKINNER
GARY MERSHAM

OXFORD
UNIVERSITY PRESS

OXFORD
UNIVERSITY PRESS

Great Clarendon Street, Oxford OX2 6DP

Oxford University Press is a department of the University of Oxford.
It furthers the University's objective of excellence in research,
scholarship,and education by publishing worldwide in

Oxford New York

Auckland Bangkok Buenos Aires Cape Town Chennai
Dar es Salaam Delhi Hong Kong Istanbul Karachi Kolkata
Kuala Lumpur Madrid Melbourne Mexico City Mumbai Nairobi
São Paulo Shanghai Taipei Tokyo Toronto

Oxford is a registered trade mark of Oxford University Press
in the UK and certain other countries

Published in South Africa
by Oxford University Press Southern Africa, Cape Town

Disaster Management 2nd edition
ISBN 0 19 578313 1

Editor: Pat Brennan
Commissioning editor: Marian Griffin
Design and cover design: Chris Davis
Proofreader: Pat Daykin

Published by Oxford University Press Southern Africa
PO Box 12119, N1 City, 7463, Cape Town, South Africa

Set in 11 on 14 pt Bodoni Book by Global Graphics
Cover reproduction by The Image Bureau
Imagesetting by Castle Graphics
Printed and bound by ABC Press, Epping Industria II, Cape Town

Contents

9 Media relations

10 Follow-up communication and performance checklist

11 Checklists

Foreword

Despite the time that has elapsed since that fateful day, many people in the world still think of the events of September 11, 2001 in the United States with incredulity. Described by someone as the mother of all disasters, the collapse of the famous twin towers of the World Trade Centre in New York City through a deliberate human act of destruction will remain indelible in our minds for many years.

There are also fires, floods, earthquakes, volcanic eruptions, hurricanes, and other natural disasters – afflicting many parts of the world at great loss in human life – to confront. Add to that derailments, aircraft and mining disasters, chemical spills, oil tanker incidents at sea, etc. Many lives would be saved each year if these disasters could be predicted and avoided or stopped before they occurred. However, we are far from that ideal state and we thus have to make do with what we can to reduce loss of life and to temper the devastation. We cannot predict the future, but we can anticipate and prepare ourselves for what might happen.

As Skinner & Mersham say, 'The challenge is to turn that "unexpectedness" into "anticipation and preparation" by application of scenario planning and analysis of broad natural and social trends. A common characteristic, however, relevant to all disasters and crises, is that decisions must be made quickly.'

Indeed, time is always of the essence in a crisis. While it is generally accepted that prevention is better than cure, we know that these unexpected occurrences will continue to happen, and as such we should learn to co-exist with them. However, to cope better we must acquire the necessary skills to deal speedily and effectively with any crisis or disaster. One skill that we must develop is to be prepared and ready with an execution plan for any eventuality. *Disaster Management* provides us with more than enough insights and tools to do exactly that: to understand what we are up against and how to respond effectively to the crisis or disaster.

This book sharpens our awareness of the fact that crises and disasters have a life and that we need to understand their life cycles if we are to tilt the scale of co-existence in our favour. We also learn very quickly that crises and disasters do not only happen to 'other' people or companies. They can happen to anyone, including 'us', at any time. Therefore, by being well prepared and knowing what to do in a crisis or disaster, we can readily identify any silver linings and even turn an apparently hopeless situation into an opportunity. Our ability to read the warning signs or 'prodromes' should increase our sense of anticipation and thus enhance our ability to respond proactively in certain cases.

As the authors guide us so ably through this well-written book, our understanding of disaster management as well as our levels of confidence should increase. This should enhance our ability to manage any crisis or disaster – including our own stress levels – with professional efficiency.

Both professional practitioners and students should find the style, register, and presentation of the book useful. There are between the covers of this book treasures to be mined, enjoyed, and exploited to good advantage.

SEJAMOTHOPO MOTAU
Ambassador at Large and former president,
Public Relations Institute of Southern Africa

Acknowledgements

Initial research and planning for this book has taken place over three years. When we started this project we never imagined that the subject matter would assume such importance, culminating in events as dramatic as September 11, 2001 in New York in the USA and subsequent natural, economic, social, and political disasters around the globe. It has, in fact, been impossible to catalogue crises as they have occurred; but this was not our intention. We simply wanted to provide a framework and planning guide for disaster management at various levels, from personal to organizational, and to highlight the importance of issues management and crisis communication.

In addition to our own research and personal in-depth interviews, we are indebted to a number of experts in the field for their encouragement and advice. Immediate Past President of the Public Relations Institute of Southern Africa (PRISA) Mr Sejamothopo Motau has been kind enough to write the foreword to share his perspectives on the subject.

Mr Joey Singh, SHEQ Manager, Umbogintwini Operation Services, gave us a 'blueprint' of the disaster management plan for one of the largest chemical complexes in the southern hemisphere.

Ms Honi Brian provided some excellent 'snapshots' to illustrate the basic text. With over 22 years experience as a PR practitioner and trainer, she currently heads up a consultancy that specializes in custom-designed crisis management programmes for high risk operations. Clients range from chemical manufacturers and others in the process and distribution industries, to healthcare, tourism, and hospitality industry players.

Ms Pat Barnes, Managing Director of BlackRock New Era Communications, also provided us with some valuable insights, including the Ellis Park soccer disaster case study which won a silver award in the Financial Mail/PRISM competition for 2001.

We are indebted too to Clive Morgan, a counselling psychologist in private practice, for his personal insights and practical advice in the field of stress management.

Other colleagues and business associates who have provided expert advice include Mr Peter Kiuluku, Public Relations and Marketing Manager of the East and Southern African Management Institute (ESAMI); Ms Mogie Subban, Senior Lecturer in the Department of Management at the Durban Institute of Technology (DIT); and Dr Derina Holtzhausen, Assistant Professor, School of Mass Communications, University of South Florida.

To all of them we express our deep gratitude and thanks. We hope through our endeavours to have shed some light on a fascinating and complex subject.

The authors
CHRIS SKINNER and GARY MERSHAM

1

Disaster management in national and global contexts

Expect the unexpected

In this book, we proceed from the paradox that while disasters and crises are for the most part 'unexpected', they can, to varying degrees, be anticipated. This proposition needs some elaboration. A disaster or crisis is a low-probability, high-impact event that threatens the viability of an organization or organizations and is characterized by ambiguity of cause, effect, and means of resolution. The attack on the World Trade Centre in New York on September 11, 2001 is perhaps the most obvious example. Most people see it as one of the most unexpected events in world history. The broad world economic downturn might, on the other hand, be seen as inevitable, if not predictable, following upon a period of astonishing growth in many major economies. Yet it was unexpected by many, and none could have predicted how the events of September 11 would act as a trigger for a period of further economic decline.

Whereas business managers and public relations practitioners typically cannot predict a *specific* disaster or crisis, they can anticipate that disasters and crises *will* happen. It is the 'unexpected' nature of each specific event that creates a crisis situation.

The challenge is to turn that 'unexpectedness' into 'anticipation and preparation' by application of scenario planning and analysis of broad natural and social trends. A common characteristic, however, relevant to all disasters and crises, is that decisions must be made quickly. Therefore conscious preparation for the inevitability of such events is required.

Natural disasters

Natural disasters are perhaps the most 'unexpected' and costly overall in terms of loss of human life and resources. In the last few years, natural disasters have claimed 100 000 lives costing 140 billion US dollars (Bendimerad 2001).

Recent natural disasters between 1995 and 2000, such as Hurricanes Mitch and George in the Caribbean and Central America, the floods in Venezuela and Mozambique, and the earthquakes in Colombia, Turkey, Taiwan and the Congo, highlight the ceaseless turmoil unleashed by the forces of nature. Experience shows that a hasty response, one that is not based on familiarity with local conditions and that does not complement the national efforts, only causes chaos. In most cases, the needs are logistical and financial, requiring the integration of the provision of support and supplies through national and international agencies.

Human disasters

At another level, we see unfolding a complex new world, driven by globalization, free markets and the communications revolution, facing long-standing and persistent problems that seem to be deepening.

In the early 1990s, an optimistic business community saw the demise of communism and the Cold War, and the opening up of international markets and free trade agreements. The days of fear of a nuclear war were over and an era of prosperity seemed to promise the improvement of the lives of the world's poor and under-developed peoples.

But poverty, hunger, health problems, widening social disparities, ethnic violence, separatism, religious intolerance and terrorism continue to be part of contemporary global society and we are increasingly aware of them through the communication media. With the onset of the revolution in digital communications and information, ignorance can no longer excuse the disparities that separate the rich and the poor nations, and the advantaged and dispossessed within those nations. It is common knowledge that over 90 per cent of global financial debt is owed by the peoples of the developing world and it is the poor of those countries who suffer as a result.

The world has shrunk and it is not possible to be isolated or immune from any part of it. In many ways, local affairs have become world affairs and vice versa. As Kofi Annan, the UN secretary-general, puts it:

> *No one today is unaware of this divide between the world's rich and poor. No one today can claim ignorance of the cost that this divide imposes on the poor and dispossessed, who are no less deserving of human dignity, fundamental freedoms, security, food and education than any of us. The cost, however, is not borne by them alone. Ultimately, it is borne by all of us...*
>
> *Today's real borders are not between nations, but between powerful and powerless, free and fettered, privileged and humiliated. Today, no walls can separate humanitarian or human rights crises in one part of the world from national security crises in another.*
>
> (Excerpt from the acceptance speech by Kofi Annan, UN secretary-general, on receiving the Nobel Peace Prize in 2001)

The free market has not alleviated poverty – the root of many crises in the world today. As Raymond Parsons (Parsons 2001) argues, the world economy has become more integrated. A downturn in one major economy such as that of the United States now spreads faster to others. The current slowdown is more widespread than in previous world recessions in 1975, 1982 and 1991. A significant difference from previous downturns is that the three main 'engines' of economic growth – the US, Europe and Japan – all weakened simultaneously.

The strong economic growth of the United States in the 1990s spilled over into many – though not all – economies, ranging from Taiwan to Ireland. As Parsons puts it, 'globalization also has a reverse gear when things go sour and now presents challenges to policy makers and business' (Parsons 2001: 1).

This combination of factors – the growing divide between rich and poor, globalization and a world economic recession – has a powerful impact on small and large businesses and major implications for the communications content and processes that drive them.

In this sea of changes, business leaders and managers have to develop the ability to predict trends and scan and interpret this new environment. They need to be able to understand and manage potential crises and disasters, and are increasingly expected to demonstrate their position on the issues involved and to take action to resolve them. The integration of world politics and global economics has led to the fall of the walls of economic protectionism, cultural insensitivity and political and social ignorance that once excused and protected business in both small and large economies.

Sources of possible crises in the near future

Specifically, what can business expect in the next few years? Control Risks Group, an international business risk consultancy, has identified eight key risks that business leaders will face in the future in its Annual Risk Survey, RiskMap 2002.

- Further terrorist attacks in developed countries likely
- Use of alternative weapons of terror, including bio-terrorism
- An increase in globalized crime, in drugs, arms, money laundering and immigrant smuggling, as a result of an economic slowdown
- Increased political violence and criminal activity fuelled by poverty and higher unemployment
- More white-collar crime
- Unforeseen disruption to business travel arrangements
- A greater reluctance among struggling economies to implement overdue economic reform programmes
- Attacks against foreign economic interests in high-risk countries, including possible kidnapping and extortion scams by various terrorist groups and freedom fighters

Corporate governance and business ethics will increasingly come under the spotlight following an extensive period of corporate accounting scandals. The activities and salaries of corporate executives will come under vigorous public scrutiny and tighter government controls will be introduced for corporate reporting.

Actions taken by countries against terrorism have resulted, in some cases, in restrictions on civil liberties. For large businesses there will be increasing scrutiny of financial transactions as governments crack down on money laundering and seek to identify organizations that give financial support to terrorist activities.

The RiskMap report also identifies six critical *issues* facing the business community in the coming year:

1 Corruption has become the preferred weapon for organized crime. This, and the blurring of the once-clear distinction between terrorism and organized crime, demands that companies must commit resources to combating corruption.

2 In a world of uncertainty, *loyalty* is no longer deemed a priority by employers or employees. Such changes in the 'worker/company' relationship lead to serious risks of malicious damage and fraud. Understanding and managing 'social capital' (employees) will be more important than ever.

3 A natural response to September 11 has been for global companies to reconsider, where possible, the necessity of international staff travel and the potential risks to operations in countries where strong anti-globalization sentiments exist. A conclusion of RiskMap is that while knee-jerk reactions are inappropriate, corporates should act to minimize risk.

4 For global businesses with overseas employees, the economic slowdown and fluctuating world situation emphasize the need for corporates to *adopt new expatriate policies*. These include balancing cost and security against the need for business control at a local level and the requirement for local skills, while at the same time maintaining overall quality and safety.

5 The international war on terrorism is likely to create and increase anti-western, anti-multinational sentiment in countries where such 'terrorists' are seen as 'freedom fighters' by sections of the populace. RiskMap highlights the need for companies to ensure that this scenario is factored into their entry strategies or continued investment plans in any particular territory.

6 The events of September 11 have demonstrated *the physical and economic vulnerability of the 'global elite'*, particularly the United States, to distant conflicts. RiskMap finds that corporates must hold firm, mitigate any risks and continue with 'business as usual'.

Source: Control RiskMap 2002.http://www.crg.com/press/release11.htm 1/8/2002

South Africa

In the Africa section of its annual RiskMap 2002 survey, Control Risks Group says South Africa and Nigeria, the continent's key countries, face a further period of awkward political choices and urgent structural reform. Control Risks Group advises 86 of the US Fortune Top 100 companies' investment decision makers. In South Africa, the pressures on President Mbeki and his government to balance fiscal prudence with demonstrable socio-economic progress will grow. South Africa leads a group of countries in a low political risk rating, although the security risk rating is medium with some areas classed as high because of the crime rate[1].

In a regional context, South Africa leads a group of countries that includes Botswana, Mauritius, Mali, Gabon and Senegal. These have low political risk, which is a positive factor for further investment.

Conflict contagion

Conflict contagion is a term that highlights the reputational risks associated with regimes engaged in civil conflict, the prominent example being the impact of the crisis in Zimbabwe on perceptions outside of and within South Africa. The South African economy has already suffered from declining investor confidence and falling trade levels associated with the deteriorating political and security situation and attacks on media freedoms in neighbouring Zimbabwe. Investors fear that an upsurge in violence and further seizures of privately owned assets in Zimbabwe could result in similar actions spreading into South Africa if landless South Africans copycat their Zimbabwean counterparts. The fact that minority white farmers still own 87 per cent of the productive agricultural land in South Africa makes this scenario more likely.

Related issues revolve around illegal immigration and refugee patterns throughout southern Africa, and political repression in Swaziland.

---

[1] RiskMap rates countries in terms of their security risks in five categories: Extreme, High, Medium, Low, and Insignificant. A 'High' security risk like the one accorded to some areas of South Africa means that there is a probability that foreign companies will face security problems and special measures are required. Assets and personnel are at constant risk from violence or theft by state or non-state actors OR there is a high risk of collateral damage from terrorism or other violence. State protection is very limited.

The effect of HIV/Aids on business

Predicting the economic outcomes of a phenomenon such as HIV/Aids should be treated with circumspection since reliable, accurate data is scarce and may be skewed when it is extrapolated into predictive economic models.

One fact is perfectly clear, however. The epidemic primarily affects young working-age adults and far exceeds any other threat to the health and well-being of the South African workforce. Over the next decade, the number of employees lost to Aids could be 40–50 per cent of the workforce (LoveLife 2001: 12). As many as 800 000 South Africans a year could be dying as a result of Aids by 2010; and this cannot fail to have a massive socio-economic effect. The epidemic will have significant direct and indirect costs.

Direct costs to companies include the costs of health care and other employee benefits. Already HIV/Aids has resulted in rising costs of employee benefits and the cost of an average set of benefits is expected to double over the next five to ten years, unless the benefits are restructured.

The most significant costs to business are likely to be indirect. These include costs of absenteeism due to illness or funeral attendance, lost skills, training and recruitment costs, and reduced work performance and lower productivity. By 2010, it is estimated that approximately 15 per cent of highly skilled employees will have contracted HIV. Labour-intensive companies may seem to be at a higher risk of lost production, but some technology and capital-intensive industries may be more vulnerable, especially those that employ specialists such as highly trained machine operators and technical staff. In small and medium-sized enterprises, the illness and death of owners or managers, or of a small pool of locally sourced skilled people, could prove disastrous.

Factors that need to be taken into account by businesses include the risk profile of employees, HIV/Aids education programmes and risk modification programmes. Businesses may face crises when key suppliers and distributors are impacted negatively. Particularly important in this area are the suppliers of water, electricity, telecommunications and government services, where breakdowns in continuity can have downstream effects. Impact assessment by forward planning for crisis management by all sectors of private and public spheres will be critical in minimizing these problems.

Impact on markets

The vulnerability levels of particular markets will be influenced by the nature of goods and services produced and the demographic and risk profile of consumers. Certain markets will expand, notably health care services and funeral services. Non-essential and luxury goods and services are more likely to be affected negatively by household expenditure shifts. Many middle income households will become poor, and goods and services aimed at upwardly mobile households may be affected. The risk of default on credit payments will increase. Long-term lenders and insurers have already begun adapting products to reduce their exposure. Savings levels and credit supply are likely to suffer further reductions.

Macro-economic impact

It is clear that the epidemic has the potential to restrict economic growth through:

- reducing the numbers of workers in the economy and thereby increasing production costs
- decreasing public sector, corporate and personal savings due to health care and related HIV/Aids expenses
- creating a risk barrier to further Direct Foreign Investment (DFI) and domestic investment
- reducing direct government investment in other areas of infrastructure development as expenditure on HIV/Aids increases

Recent estimates have suggested that HIV/Aids could reduce Gross Domestic Product (GDP) growth rates by between 0,3 and 0,4 per cent per annum over the next 15 years. The impact on human and social development is expected to be much more profound than reflected by limited indicators such as GDP or per capita GDP. Affected persons, particularly orphans, will also have greatly reduced chances of fulfilling their human potential. Reduced parental care may also increase antisocial behaviour, crime levels and the number of homeless, and will decrease educational opportunities. Finally, HIV/Aids will increase socio-economic disparities (LoveLife 2001).

Crime

South Africa is internationally notorious for high rates of violent crime. Johannesburg has a reputation as one of the crime centres of the world. The estimated murder rate of 55,3 per 100 000 people remains one of the world's highest and the threat of crime is likely to remain serious over the medium term. Ease of access by criminals to weapons is a legacy of past and current conflict, both in South Africa and in the sub-region. Jobless youth often turn to crime, and the high unemployment rate continues to increase. Xenophobia is also on the increase as unemployed South Africans perceive immigrants as a threat, acquiring jobs at their expense or more successfully engaging in entrepreneurial activities. Finally, international drug-smuggling cartels are increasingly making use of South Africa as a transit corridor, and using South Africans as accomplices.

Potential religious and cultural conflict

Renewed religious tensions involving a significant Islamic population remain a potential trigger for violence. The Islamic religion in South Africa is associated with people of Asian descent, and according to the South African Human Rights Commission, adherents of Islam have been stigmatized by association with Arab terrorism (SAHRC 2000: 87). The Media Review Network has complained that followers of Islam in South Africa are unfairly stereotyped in the media as being prone to terrorist activities. They argue that South African media consistently stereotype adherents of the Muslim religion by association with violence or terrorism (SAHRC 2000: 21, 86).

Punitive US military action against Muslim countries suspected of harbouring terrorist cells linked to the September 11 attacks in the US might provoke protest marches and maybe also a resurgence of a bombing campaign against Western commercial targets and government offices.

In addition, dissension within the Muslim community over such issues has led to violence in the past. A *Cape Argus* article suggested that the motive for the bombing of Dr Ebrahim Moosa, an internationally respected authority on Islam, might have been retribution for his criticism of People Against Gangsterism and Drugs (PAGAD) (SAHRC 2000: 46).

Implications for the business community are that non-Muslim employers should become more familiar with the Islamic culture and religion, the role it plays in social and community stability and become sensitive to Islamic practices and the Islamic dress code to ensure religious accommodation in the workplace.

Conclusion

Perhaps the greatest threat to and enemy of business in the coming years is fanaticism, fuelled by lack of education, poverty and relative deprivation. From the stone-throwing Palestinian youths on the West Bank to the Timothy McVeigh race-inspired bombings in Oklahoma City, this point is borne out. It is brought into sharp relief by the contrast between Muslims who burnt effigies of Bush and Blair and by those who stood up and said this is not the way of Islam. It is clear that the most important issue at stake is cultural understanding. As organizations, we need to understand the broad social trends that are challenging and reinventing our taken-for-granted traditions, ideologies and values.

World leaders such as Tony Blair are attempting to widen the debate over the rebuilding of Afghanistan to include the acknowledgement of the factors that lead to poor and developing regions of the world becoming fertile breeding grounds for terrorism. His statements on Africa in this connection are relevant, describing the condition of the peoples of Africa as a 'scar which can be healed'.

As Martin Prozesky (2001) argues, the business world is the new epicentre of ethical creativity. In the contemporary world, business leaders are beginning to realize the logic of the successful pursuit of profit and of effective morality, 'making nonsense of the old joke that business and ethics is a contradiction in terms'.

We are by nature heading irrevocably towards the global village (some would argue global suburb) driven by global partnerships. This reality generates a crucial choice. We can either accept the reality of globalization and work to win the willing cooperation of others, or we can attempt to force and manipulate others. We have the unsuccessful results of the second option under apartheid in South Africa and under communism in Eastern Europe. It failed in a world where most people were poorly educated, isolated and economically disempowered. How can it possibly succeed in a world where more and more people are connected, aware and able to mobilize mass action through the global power of the Internet, as witnessed at Seattle in 1999? That leaves just one option – except for those unwise enough to be interested only in short-term gains for their organizations – and that is the option of finding ways in which human creativity, energy, and resources can work together in what Martin Prozesky calls 'willing partnerships'.

Sources

Bendimerad, F (2001) *Megacities, Megarisk*, [Online], Available: http://www.worldbank.org/html/fpd/dmf/megacities.htm [accessed 08/01/2002].

Control RiskMap (2002) [Online], Available: http://www.crg.com/press/release11.htm [accessed 1/8/2002].

De Ville de Goyet, C (2001) *Myths and Realities* [Online], Available: http://www.worldbank.org/html/fpd/dmf/myths.htm [accessed 08/01/2002].

Global Alliance for Public Relations and Communications Management Association (October 2000) *21st Century issues and trends – the views of corporate executives in the United Kingdom.*

LoveLife (2001) *The impending catastrophe: a resource book on the emerging HIV/Aids epidemic in South Africa*, Johannesburg: Abt Associates.

Media Tenor (2001) Mbeki not given appropriate platform? International television news programmes coverage of their heads of state, Vol 2(4), Pretoria: Institute of Media Analysis.

Mersham, GM & Skinner, JC (1999) *New Insights into Communication and Public Relations*, Johannesburg: Heinemann Higher Education.

Mersham, GM & Skinner, JC (2001) *New Insights into Business and Organizational Communication*, Johannesburg: Heinemann Higher Education.

Mersham, GM & Skinner, JC (2001) *New Insights into the Communications and Media*, Johannesburg: Heinemann Higher Education.

Parsons, R (2001) 'The global and domestic economic outlook', *Durban Chamber of Business Digest*, December, issue 23.

Prozesky, M (2001) 'The new epicentre of ethics', *The Natal Witness*, 27 February.

South African Human Rights Commission (SAHRC) (2000) 'Faultlines: inquiry into racism in the media', *SAHRC Report*, August 2000.

Stovin-Bradford, R (2001) 'Hard-hitting report takes a stern view of SA', in *Business Times*, 11 November: 8.

2

Disasters, issues and crisis management – the similarities and differences

Definitions

It is important to define the key terms used throughout this book and indicate how they are interrelated.

- A *crisis* is an unstable or crucial time or state of affairs whose outcome will make a decisive difference, for better or worse, for an organization.
- A *disaster* may be defined as a rapid disruption of routine operations causing serious damage to property and/or injury to people.
- An *issue* may be defined as an unresolved or ambiguous social or policy matter of public concern, which, if acted upon by particular groups, may have either an adverse or a positive effect on the business.
- Ultimately a disaster or an unresolved issue may lead to a *crisis* – when a particular state of affairs reaches a climax that leads to often-unexpected consequences. Research shows that complacency – self-satisfaction accompanied by unawareness of actual dangers or deficiencies – is commonly an element present prior to a crisis reaching its highest point. Often, it is at this peak that a situation is characterized by being 'out of control', and outside forces and unintended and unexpected consequences overwhelm management's ability to cope.

The following diagram reflects the scope of disaster management:

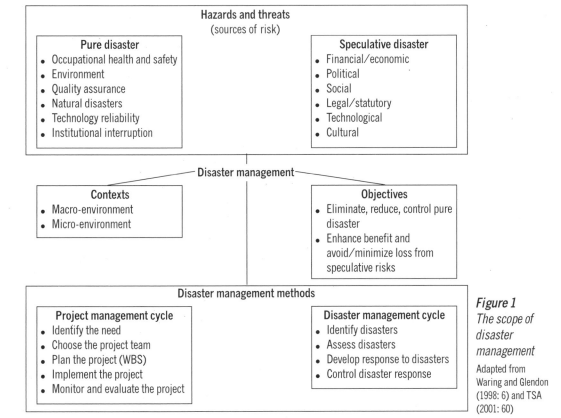

Figure 1
The scope of disaster management
Adapted from Waring and Glendon (1998: 6) and TSA (2001: 60)

The sources of risk can be divided into pure disasters and speculative disasters. In turn, disaster management needs to be seen in the context of the macro and micro environment, with clear objectives of taking control and minimizing or avoiding loss.

A multi-dimensional approach can be used to manage disasters, but it must be borne in mind that the management of disasters should essentially be an ongoing process and part of the greater plan of any department or division.

Issues management

What business is able to do increasingly depends on what it is allowed to do, both by legislation and public opinion. Consequently, the central challenge is for companies to identify future issues and respond with timely strategies and programmes.

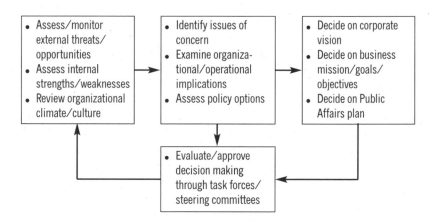

Issues management therefore is an 'early warning' process that enables management to:

- be better informed about the changes in the business environment and be more sensitive to the public implications of business decisions
- demonstrate to stakeholders an understanding of the organization's business and its impact upon society (often simply taken for granted)
- anticipate legislative and regulatory issues based upon emerging and future public policy and corporate issues
- lead business and industry initiatives
- become more active in external affairs and shape the political development of issues and the outside view of the company
- avoid 'surprises' which cost money and time and put the company's reputation at risk

Issues, as we will see later in the book, can also provide key marketing opportunities.

Typically, the corporate response to issues management is reactive rather than proactive. As Figure 3 shows, it is usually directly proportional to external pressure, and therefore the organization is seen publicly as being controlled by external factors as it endeavours to carry out damage control. Improved goodwill and positive image would accrue from a proactive approach where the organization is seen to be leading the process in consultation with other stakeholders such as government and pressure groups.

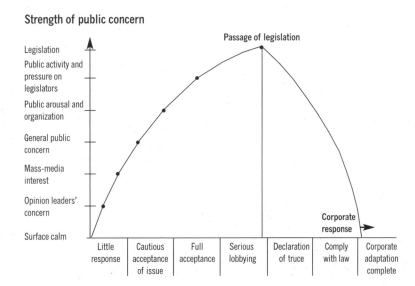

Figure 3
*Corporate response
to public concern*

Who should be involved in issues management?

It needs to be made clear here that issues management, guided by
the issues management team, should be a permanent structure
within the organization. It should not be confused with the forma-
tion of a crisis communication management team, although some
of the members will be common to both; for example, during a crisis period, the team may be larger including more specialized personnel in media relations and logistics, as well as external consultants. The emphasis in issues management is upon trend analysis
and proactive steps to obviate the onset of a crisis.

At the executive level, it should be the responsibility of the
CEO and the marketing, corporate affairs, technical, human
resource, information technology and legal directors/public
affairs advisers to identify potential issues. At the managerial
level, issues that could impact on day-to-day activities must be
brought to the attention of the executive team. The communication specialists within the group such as the public relations,
media and internal communication managers, as well as outside
consultants, should also participate.

Seven steps of the issues management process are shown below
in Figure 4:

The issues management process

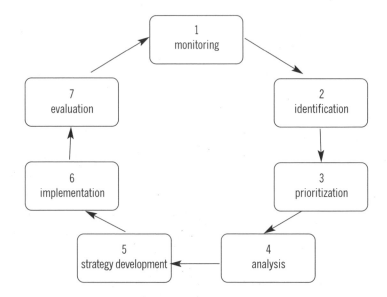

Figure 4
The issues
management
process

1 Monitor the business environment

The first step is to introduce a comprehensive environmental scanning programme to systematically monitor any political, economic, social developmental and cultural trends that may affect the company or its divisions. At the same time, establish regular contacts with key sources of information including: media, interest groups, governments, opinion leaders, trade unions, religious organizations, academics and external research services.

2 Identify important issues

All possible issues that are likely to affect business objectives or activities should be identified from the information gathered in the scanning exercise and from personal contacts. This is a useful exercise that puts on the table many issues and threats that otherwise lurk below the surface.

3 Prioritize key issues

In any multinational operation, any one company in the group may have many issues to consider, so allocate appropriate time and resources by classifying issues in order of priority. Assess an issue's probability of occurrence and impact on the product, sector, company or industry. Set up a small task force or work group with those line managers who are or should be concerned about the issue.

4   Select and analyse key issues

The work group should submit a report to management containing background and current status and likely development of the identified issues, possible policy options open to the company and recommended strategy and supporting public relations programme.

5   Ask management to decide on strategy

Management ultimately has to decide on whether or not to approve the work group's report. Strategy based on the report might include the company's response, target groups, timing and resources needed to execute a programme.

6   Implement the programme

The public relations team and the business sectors involved should be responsible for implementing the programme chosen by the management team. Their aim should be to establish and strengthen contacts with target groups and communicate the company's position, both internally and externally, in a professional and credible manner. This is crucial at this stage.

7   Evaluate success and failures

The company should try to measure the success of its strategy and programmes so that it can be better prepared for future events, but above all, to improve communication with the key stakeholders.

Comparisons between issues and crisis management

On the whole, the principles and disciplines of crisis management work equally well for issues because the two are so closely related. For example, a major health or environmental concern highlighted by a pressure group and/or the media, which is in essence an issue, can threaten a whole company's or industry's viability to the extent that it becomes a crisis. So, to a large extent, there will be an overlap in terms of the personnel who are involved in both issues and crisis management. However, there are differences between crisis and issues management identified by Bland (1998: 172–177) which indicate how different personnel and time frames are involved.

> *A crisis is an issue in a hurry. You could also say that an issue is an infant crisis. Both present some kind of threat – to your reputation, your bottom line, your licence to operate and so on – but usually over a different period of time. Crisis management therefore requires more in terms of advance planning, team building, training, simulations and prepared plans. You have to be able to 'press a button' and preplanned operations rapidly fall into place.*
>
> (Bland, 1998: 172–177)

By contrast, Bland adds that hurriedly pressing a button may be the last thing you should do when handling an issue. As long as your early-warning system is working and you have identified the issues developing, it is usually best to evolve a response, starting with brain-storming groups and building up to a specifically targeted programme of the right intensity for that particular issue, involving the right people. The appropriate team training and planning will present themselves as the issue is addressed.

Essentially, both crisis and issues management involve public relations-led techniques to protect the company's reputation and/or licence to operate when under threat from negative outside influences.

So if we take the crisis preparation and handling checklists, and apply them to issues management, we end up with very similar lists. The common ground is:

Preparation

- **What** issues could hit us?
- Who are the **audiences**?
- How do we **communicate** with them?
- What are the **messages**?
- What are the **resources** and **facilities**?
- Are spokespeople **trained**?
- Have we built **bridges** with our audiences?

Handling

- **Assess** the situation.
- Select and assemble the **communications team**.

- Decide on the **strategy** and **plan**.
- Identify the **audiences**.
- Decide on the **messages**.
- **Brief** relevant people.
- **Centralize** information.
- **Understand** your audiences.
- Give **information**.
- Give **reassurance**.
- **Resist** combat.
- Be **flexible**.
- Think **long-term**.

Bland argues that the issues communications team will be less structured and more flexible and changeable than for a crisis, but it will still involve a number of the same people.

> *As with successful crisis management, it is not just about communicating – it is more about how you communicate. As ever, success comes from understanding your audiences, being seen to give some ground and demonstrating that you see things from their point of view before getting on to your own agenda.*
>
> (Bland, 1998: 177)

It is well to remember that issues, like crises, can also present opportunities as well as threats to companies. Use the public spotlight to demonstrate the caring and responsible aspects of the company under adverse circumstances. This is often not possible under so-called normal conditions when good news simply falls on deaf ears!

Snapshot: Initiatives in progressing crisis preparedness

Significant initiatives have emerged during the past decade to provide companies and the communities within which they operate with guidelines on crisis management. These have generally come from bodies and institutions that represent specific interest groups or industries.

The United Nations response to environmental effects: a model for community preparedness

Foremost amongst these is the APELL (Awareness and Preparedness for Emergencies at Local Level) process. Developed in 1986 by the Industry and

Environment Office of UNEP (United Nations Environment Programme), in collaboration with the industrial sector, it formed a response to the adverse environmental impact of a number of industrial accidents. Amongst the events that precipitated APELL were industrial disasters such as:

- a dioxin release in Seveso, Italy in 1976
- a propane explosion in Mexico City in 1984
- the release of methylisocyanate at Bhopal in 1984
- a fire in a warehouse in Basle and subsequent contamination of the Rhine in the mid 1980s

as well as such natural disasters as

- the earthquake that struck Mexico City in 1985
- mud slides in Ecuador in 1987
- the eruption of Mount Nyiragongo, near Goma in the Democratic Republic of Congo in 2002.

The rationale for APELL is universal acknowledgement that, whatever its cause, every disaster has an environmental impact. While the process was developed as a means of responding to technological accidents, its underlying philosophy addresses issues of safety and emergency preparedness for all peoples in all countries of the world. It suggests measures that can assist governments to minimize the harmful impact of chemical accidents and industrial events that could result in environmental damage and loss of life and property.

The APELL process encourages the view that, while accidents are preventable, it is advisable to prepare emergency response plans. These are of necessity based on an understanding of local hazards. This understanding can also help in developing measures to prevent the occurrence of a disaster.

In constructing the APELL process, UNEP acknowledged that culture, value systems, infrastructure and resources, and regulatory requirements differ from community to community and between countries, and that different potential dangers and risks are found in different industries. But it considered one need common to all situations: the ability to cope with an industrial disaster with the potential to affect the surrounding community. APELL accordingly provides basic concepts and guidelines that focus on awareness and preparedness.

Based on cooperative action between local authorities, community leaders, industry managers and other local interested parties and stakeholders, APELL has two distinct yet interlinked components:

- Community Awareness, which is achieved through providing information about possible hazards resulting from the manufacture, handling, and use of hazardous materials, and steps taken by the authorities and by industry to protect the community; and
- Emergency Response, which relies on the formulation of a plan, based on the above information and developed in cooperation with the local community, to protect the entire community should a situation arise that endangers its safety.

The APELL focus is local, since it is generally acknowledged that the initial response to an incident influences both the final outcome and its magnitude, and that this initial response is usually provided locally.

Relevant national and international involvement is, however, also essential. UNEP therefore advocates that local industries and communities integrate their efforts with the programmes of those authorities, departments, agencies and others responsible for national planning, for industry and the environment, and for public services, as well as for disaster management, safety and security. Broad-based cooperation of this nature is a requisite to ensure the coordinated effort that will be necessary should national or even international resources have to be mobilized in the event that an incident escalates to proportions that cannot be dealt with at local level alone.

An APELL process handbook gives comprehensive guidelines. It details the roles and responsibilities of the various partners – national governments, industrial facility owners and managers, local authorities and community leaders – and of a coordinating group that represents the interests of these stakeholders. This group needs to be highly organized, as it represents the broad variety of functions and agencies that will be brought into play in the event of a disaster.

The handbook spells out basic principles, such as the needs and rights of local communities to be informed about hazardous installations, as embodied in the *Guidelines for World Industry* formulated by the International Chamber of Commerce, and proposes actions for plant managers, local authorities and community leaders to take individually or within the coordinating group, to build awareness of industrial activities, and to ensure that information exchange is an effective two-way process.

A myriad of issues that have the potential to precipitate a disaster on the one hand, or to hinder or help an emergency response process on the other, needs to be identified, assessed and addressed as a preliminary step. These issues provide the basis for the design of a viable, coordinated emergency response mechanism for the community, which the APELL model represents in flow chart format.

In addition to suggested timelines for implementation of the final emergency plan, the APELL process also reinforces the necessity for annual and on-going reviewing, assessment and updating, as well as the need for all organizations and agencies concerned to develop standard operating procedures that coincide with the provisions of the plan.

Of particular interest is the model's emphasis on communication, and information provision and exchange. This is reinforced at each step of the process. Communication not only provides the means whereby communities become aware of hazards and are trained in emergency response, but it also forms the medium for mutual consultation which leads to an integrated emergency plan that is acceptable to all stakeholders, resulting in the higher level of 'buy-in' to the coordinated and collaborative effort required of community participants should a problem arise.

Lastly, the model is a powerful tool that can enhance or diminish the reputations of companies and agencies who have to deal with a crisis situation. The way communication is managed, and its content, relevance and timeliness, both during a crisis and during the post-emergency period, can have a lasting impact on public and community perceptions through the impressions that are created – either of effectiveness and transparency, or, as is still too frequently the case, the opposite.

Information supplied by Honi Brian, Cormark Communication, cormark@ircradio.co.za

Sources

Bland, M (1998) *Communicating out of a Crisis*. London: Macmillan Business.

Cutlip, SM, Center, AM & Broom, GM, (2000) *Effective Public Relations* New Jersey: Prentice Hall.

Fearn-Banks, K (1996) *Crisis Communication: A Casebook Approach*, New Jersey: Lawrence Erlbaum.

Seitel, FP (2001) *The Practice of Public Relations*, New Jersey: Prentice Hall.

Shell, *Public Affairs in Practice*, Section on Issues Management, London: Shell.

Skinner, JC, Von Essen, LM & Mersham, GM (2001) *Handbook of Public Relations*, sixth edition, Cape Town: OUP.

United Nations Environment Programme (1988) *APELL: Awareness and Preparedness for Emergencies at Local Level*, Paris: UNEP.

Waring, A & Glendon, AL (1998) *Managing risk: Critical issues for survival and success into the 21st Century*. London: International Thomson Business Press.

3

Identifying potential crisis situations

Crisis planning

The September 11 attack by terrorists on America demonstrated that some organizations in the United States were well prepared, and most were not. Less than five per cent of US companies have a crisis plan before having a significant crisis (Bernstein 2002). A severely neglected aspect of crisis communication is crisis prevention. Prior to suffering their first major crisis, few organizations invest the time necessary to take a hard look at their own vulnerabilities except in the context of legally required risk management. Although few organizations would logically plan to respond to a massive terrorist attack, in any field, there is no such thing as a business in which crises do not occur. What are your chances of having to deal with such situations? The answer is 'exceptionally high'.

What is required is a crisis communication plan which provides a *system* for *co-ordinated*, prompt, honest, informative and concerned responses to crises. This plan consists not only of a manual with scenarios and instructions, but also involves a comprehensive *audit* of the organization's vulnerabilities that results in numerous recommendations for operational/system changes which, if unchanged, create a potential for crises.

Commonly, the board of directors or administrative staff are aware of system weaknesses, but have not considered carefully the marketing communication/bottom-line impact of failure to quickly correct the problems.

Prevention, then, versus reaction, is the ultimate key to successful crisis communication. When a crisis plan is implemented after a damaging crisis, considerable time and expense is wasted attempting to minimize damage. Such expensive 'fire fighting' would have been unnecessary had a plan been in place.

Defining a crisis

Crises are unplanned events that directly or potentially threaten a company's reputation; the environment; the health, safety or welfare of employees; and the health, safety or welfare of citizens in communities surrounding a plant.

These include:

- natural disasters
- widespread illness
- fires, explosions, hazardous material spills, bomb threats, civil disturbances
- service interruptions
- death or serious injury due to questionable circumstances
- events which traumatize employees
- deaths of key employees
- sudden loss of key suppliers or contractors
- hostile take-overs and acquisitions

The benefits of a disaster and crisis management plan include preventing crises before they happen, dramatic enhancement of crisis response time, correction of operational weaknesses and overall reduced cost of crises.

The anatomy of a crisis

Crisis management is a process of strategic planning for a crisis or negative turning point, a process that removes some of the risk and uncertainty from the negative occurrence and thereby allows the organization to be in greater control of its own destiny.

Crisis communication is the communication between the organization and its publics prior to, during, and after the negative occurrence. The communication is designed to minimize damage to the image of the organization.

Effective crisis management includes crisis communication that not only can alleviate or eliminate the crisis, but can sometimes bring the organization a more positive reputation than before the crisis. (Fearn-Banks 1996: 2)

A crisis has five stages:

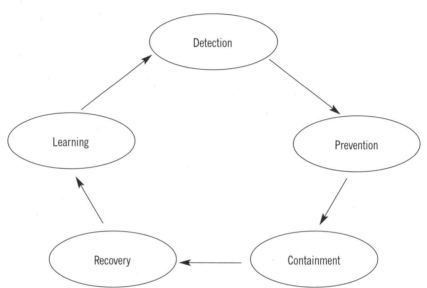

Figure 5
Anatomy of a crisis

Detection

This is a necessary and healthy exercise that puts on the table many issues and threats that often lurk below the surface because no one is sure about the correct forum to deal with them. Some companies prefer not to use the term 'crisis' because of a mindset that is in denial of 'bad news' or because they have an illogical fear of 'stirring up trouble'. Using euphemisms such as 'potentially embarrassing situations' or similar phrases is not helpful because your employees will learn that 'crisis' doesn't necessarily mean 'bad news' and 'panic', but simply 'very important to our company, act quickly'.

The detection phase may begin with noting the warning signs, what Barton (1993) referred to as prodromes or the prodromal stage. Some crises have no noticeable prodromes, but many do.

When an organization in the same business as yours suffers a crisis, it is a warning to your organization. The 1982 Tylenol tampering case was a prodrome to other manufacturers of over-the-counter drugs. Most companies heeded that warning and now use tamper-proof containers. However, in the case of Johnson & Johnson, the manufacturer of Tylenol, no one had ever poisoned their over-the-counter painkillers so it was not a crisis they had anticipated.

An organization should watch for prodromes and make attempts to stop a crisis at that stage before it develops into a full-blown crisis.

Crisis detection also refers to a system within the organization in which key personnel are immediately notified of a crisis. An organization has a considerable advantage if it knows about a crisis before its publics do, especially before the news media get the tip. It gives the organization time to draft a statement, make preparations for a news conference, notify the crisis team and call spokespersons.

Bernstein refers to the methodology of the 'vulnerability audit' in crisis prevention which entails the following:

1 Collect data from people in key information flow positions. Senior executives are not always aware of all of the circumstances which can lead to the birth of a crisis. Hence, interviews are conducted with both white- and blue-collar personnel at various echelons of the company (typically a minimum of 20 interviews). Multi-location businesses usually require interviews with remote location personnel who have insights specific to their area.

2 These interviews are conducted on an extremely confidential basis. Ideally, interviewees are told that the firm's senior management will not, under any circumstances, be told 'who said what'. Information gleaned during the interview process includes:

- potentially harmful trends (facts or perceptions reported by multiple sources)
- significant inconsistencies between answers from different subjects
- non-verbal cues that there may be something amiss in certain areas
- which then prompts further questioning; and

- consensus opinion regarding the probability of certain types of crises

3 Look for operational and communication weaknesses which could cause or contribute to a crisis. An employee who is a 'loose cannon' is a more obvious potential source of problems, even if he/she is well intentioned, but there are less obvious issues revealed through the vulnerability audit process. For example, relying on a single communication channel such as fixed line telephony for incoming and outgoing communication from headquarters offices during a crisis could be dangerous when there is a trunk line failure. The same applies to e-mail systems. The simple addition of fax machines, cellular short messaging service (sms) lists, and similar tactics can often greatly improve crisis response.

4 Report results. The conclusions from the vulnerability audit are then analysed and presented both as an in-person briefing and in writing.

5 Make recommendations for systems revisions. If there are changes (such as the aforementioned addition of fax machines) which can optimize crisis prevention and response, they are recommended.

6 Discuss scenarios most likely to affect the client company. The audit will lead to a list of 'most likely' scenarios with which the client company may deal in the future. At the in-person presentation of audit results, that list is finalized (which often results in deletion or addition of some scenarios) and then the management team brainstorms both general and audience-specific key messages for each scenario.

The information collected during the vulnerability audit process is used as the basis for writing a manual which will guide the entire organization in the communication aspects of responding to crisis situations, to include clear delineation of individual responsibilities and draft responses which reflect the company's values while considering the public's sensitivities and need to know.

Snapshot: Using the 'secret shopping' method to prevent crises

The concept of 'secret shoppers' has uses far beyond the corridors and cash registers of retail stores. Retailers and wise businesses that are highly focused on customer service have long employed people to secretly 'shop' as if they were actual customers or clients, and then report their perceptions to management. If you apply this concept to testing how an organization performs in multiple categories – not just customer service – you will be able to detect the seeds of budding crises well in advance of serious damage being caused.

Secret shopping can be done to evaluate vulnerabilities in the following areas:

- Physical and Information Security. How easy is it to just walk into a facility unchallenged? To see the contents of files containing what should be confidential information? Would it have been easy to just pick up a computer disk or CD-ROM or computer password off someone's desk? Are valuable products placed in a manner that would allow someone to easily pick them up and stick them in a pocket or purse? Are people talking about company business within easy earshot of a visitor?

- Human Resources. Does the 'secret shopper' prospective employee get treated in a manner consistent with labour law requirements and other applicable laws? Are there signs of discrim-ination or harassment in the manner people talk to each other? Are there vulnerabilities indicated in the answer the customer receives when asking any average employee, 'what's it like to work here?'

- Financial/Business Matters. Contact some vendors for the organization, playing the role of someone who's also been asked to be a vendor. Find out how they treat outside vendors (a prime source of potentially damaging gossip), whether they pay on time, and what the vendor thinks of their business practices.

- Investment Matters. If the target organization is publicly held, become a potentially large investor who wants feedback from major brokers and/or analysts. And then ask the same questions of the organization's Chief Financial Officer. Negative feedback from the former, or inconsistencies between the answers given by those outside and inside the company, are possible warning flags.

Source: Bernstein 2002.

With a little thought, you can probably see how the secret shopper concept can be extended to test much of any organization's operation, although it is not a substitute for the depth of information that can be obtained through a comprehensive vulnerability/risk assessment, discussed in this chapter.

Prevention

Continuous, on-going public relations programmes and regular two-way communication build relationships with key publics and thereby prevent crises, lessen the blows of crises, or limit the duration of crises.

Developing the right corporate culture within the organization, encouraging open interaction of members and including crisis management in the strategic planning process also minimizes crises.

Providing prompt responses to media inquiries and, if necessary, consulting respected experts on issues to refute charges can often defuse possible crises. There are crises that cannot be prevented, but every company should have a crisis communication plan, telling each key person on the crisis team what his or her role is, whom to notify, how to reach people, what to say and so on.

Containment

Containment refers to the effort to limit the duration of the crisis or keep it from spreading to other areas affecting the organization.

Recovery

Recovery refers to efforts to return the company to business as usual. Organizations will want to leave the crisis behind and restore normalcy as soon as possible. It may mean restoring the confidence of key publics, which means communicating this return to normal business.

Learning

The learning phase is a process of examining the crisis and determining what was lost, what was gained, and how the organization performed in the crisis. It is an evaluative procedure also designed to make a crisis a prodrome for the future.

Crisis communication

The essential role of crisis communication is to affect the public opinion process and to be instrumental in establishing and communicating proof that the prevailing 'truth' is not factual or not wholly factual. It is a fact that in the court of public opinion, a person or organization is guilty until proven innocent. This is the reverse of experience in a court of law where a person is innocent until proven guilty.

Public opinion is difficult to define, but is based on attitudes of individuals towards specific issues. These attitudes are based on age, educational level, religion, country, state, city, neighbourhood, family background, and traditions, social class, or racial background. All of these help to form each individual's attitudes and a predominance of similar attitudes makes up public opinion.

The news media is a prime tool for changing public opinion. It can reach the masses in a short period of time, locally, nationally and internationally. Public relations personnel are trained in knowing how to reach the media, when and how to call a news conference, when and how to do one-on-one interviews, and when and how to disseminate written material.

Crisis communication, like public relations, is not merely the distribution of news releases. Neither is it only media relations. Frequently community relations, consumer relations, employee relations, investor relations, government relations, as well as many other fields of public relations are involved.

Communication objective

In the event of a minor or major incident, every effort should be made to communicate appropriately to employees, management, surrounding communities, other target publics, and the news media promptly and accurately. Specially designated communication staff will be the primary information source available to the news media.

When a crisis occurs, it is necessary for communication personnel to gather facts and data quickly, including the nature of a company's response to the crisis. The following communication efforts will strive to alleviate employees' concerns, minimize speculation by the media, and to ensure that a company's position is presented fairly and accurately.

It should be noted, though, that many incidents occur that are relatively minor in nature and are consequently not covered by the media. Nevertheless, it is essential to gather facts about the incidents and have them available to communicate, if necessary, to appropriate audiences.

The nature of that communication is outlined in the following plan.

Roles and responsibilities

Each site's Crisis Committee (at the various sites) is responsible for the following: identifying, confirming, investigating crises; developing strategies for managing crises; and developing strategies for recovering from crisis incidents.

As a member of the Crisis Committee, communication personnel will:

- provide a representative at the Emergency Operations Centre (EOC), if activated
- control the release of information to employees, surrounding communities, and to the news media
- maintain contact with media representatives
- establish and maintain a news conference centre, if necessary

Two, and in some cases three, communicators may be required for adequate crisis communication response. Those communicators will function in the following roles: Public Relations (PR) Lead, Incident Command (IC) Interface, and PR Back-up.

The responsibilities for each communication role are outlined as follows:

PR Lead

- receive initial notification
- designate staff member as IC Interface
- designate staff member as PR Back-up (if necessary)
- receive initial facts and updates from IC Interface
- prepare initial statement for release
- field media inquiries to office
- brief executive identified as spokesperson
- join EOC team, if activated
- provide updates to senior executives

IC Interface

- join IC team
- gather and document facts as they become known
- share initial facts and updates with PR lead in main office
- prepare initial statement for release (if necessary)
- brief affected organizations' management and employees
- field media inquiries (if necessary)

PR Back-up

- update corporate personnel as necessary
- field additional media inquiries to office
- field employee inquiries to office
- conduct in-field interviews with media (if necessary)
- set up media conference room (if necessary)

Audiences during a crisis

To communicate a message effectively it is essential to understand who your audience is and how you want your audience to react.

There are two key types of audiences during an emergency:
- people directly affected by the emergency
- people whose attitudes about the company might be influenced by information about the emergency

These two types of audience are broken into a number of categories. Public relations' objectives in dealing with each of these audiences are listed below:

Employees

How many spokespersons does your company have? The answer is 'the total number of employees'. Certainly an organization should have a policy whereby only certain individuals are officially authorized to speak for the record. If a journalist calls and there is a designated spokesperson policy, the call will probably be routed correctly. But the majority of communication will be unstructured, casual and 'off the record'. We can expect a secretary, an intern or a junior executive to give their version of the facts to colleagues, family members, friends, school management board members, golfing associates and anyone else they know.

We want employees to know that their safety is the number one priority during an emergency. Employees need information regarding the emergency as soon as possible. This must be accomplished in a manner that assures employees that the company has their best interests at heart and that it can effectively handle emergencies. Internal crisis communication is a vital aspect of disaster management.

Internal audiences are as, if not more, important than external audiences during a crisis, and yet those who aren't actually on the crisis response team often receive the least consideration when a crisis erupts. It is vital, during the crisis communication planning process, to formulate key messages not only for employees, but also for others who are close enough to the organization to be considered 'internal' – e.g., regular consultants and major vendors. They're the ones who are going to be asked first, by external audiences (including reporters, when they try to go around you), 'what's going on?'

Brief all employees in person about what's happening and keep them informed on a regular basis. In-person briefings say 'we care about you' in a manner which no memo, e-mail or internal newsletter can accomplish. You do not want internal audiences to read facts, or alleged facts, in your local newspaper first!

Identify your best 'unofficial spokespersons' and your 'loose cannons'. The former are employees who you know are loyal, know when to speak and who are admired by their peers; if they feel that they're receiving accurate information and are being cared for, they'll pass that feeling on to others along with the key messages you've shared. Loose cannons are those who just don't know when to shut up, whose feelings – sometimes disloyal/disgruntled, sometimes zealously loyal – lead them to communicate not only facts, but rumours and innuendo. During crises, loose cannons need to receive gentle, but firm extra counselling about appropriate communication and/or be particularly well isolated from sensitive information.

Create a rumour-control system. Provide means by which internal audiences can ask questions and get rapid responses. You can designate certain trusted individuals (white- and blue-collar) as 'rumour control reps' who will field questions and then obtain answers from someone on the official crisis response team. And it's important to also have an anonymous means of asking questions, through the 'rumour control reps' or through a centrally situated locked drop box. All employees can be encouraged to use either communication method without fear of reprisal.

Successful implementation of an internal communication programme will carry your key message better, longer and farther than most external communication, while a lack of internal communication can completely undermine even the best external strategy. The two can, and must, go hand-in-hand.

Community residents

We want to quell any unnecessary fears. We want the surrounding community residents to know that we take quick, effective steps to protect the health and welfare of community residents and the environment. This can be best accomplished by responding quickly to community concerns and need for information.

Top management

This group needs to be kept informed in the event of an emergency as well as be accessible as a resource, if necessary. Ignore any tendency to 'keep the top brass out of it until it's really necessary'. In many cases that's too late. The unhappy result of top management's lack of involvement is clearly demonstrated by the Starbucks case study below.

Snapshot: 'The Starbucks effect: When a good coffee company brews bitter feelings'

Few would argue against a company's bad practice being exposed to the public, but what about when embarrassing practices are aberrations, not the norm? An unfortunate incident occurred involving a Starbucks New York City store and ambulance workers shortly after the September 11 attacks on the World Trade Centre which tarnished Starbucks' otherwise solid reputation.

Starbucks is generally regarded as a solid corporate citizen and supporter of coffee pickers, the environment, the communities located around their stores, and the kids in inner city schools. But that did-

n't inoculate them from a huge amount of negative publicity. On September 11, the Midwood ambulance team stopped at this particular Starbucks shop in search of water to treat shock victims. Midwood's President Al Rapisarda said his employees paid cash out of their own pockets to meet the store's demand for $130 for three cases of water.

After the event, the ambulance company called Starbucks' customer service operators and explained what had happened. At that point, reports say, they weren't so much challenging having to pay, but rather checking if they were overcharged. But Midwood says

Starbucks said the incident couldn't have happened and, essentially, brushed them off.

Subsequently, the incident was described in various e-mails on the Internet. The exact chronology of events at this point becomes blurred. The company wrote a letter to Orin Smith, Starbucks' president, which wasn't answered. Eventually the e-mail attracted the attention of journalists. They contacted both Smith and Howard Schultz, the company's chairman and chief global strategist. Again, no response.

On September 25, two weeks after the attacks, a Seattle journalist wrote about the incident. What resulted is every company's nightmare. Fuelled by an Associated Press story that went worldwide, newspapers, broadcasters and websites across the globe picked up the story. Here is a sampling of headlines and responses:

- As if $3.50 for a Cappuccino Wasn't Bad Enough (website headline accompanying a call for a worldwide boycott of the company)
- Starbucks Charged Rescue Workers at the World Trade Center Collapse for Bottled Water
- Starbucks to WTC rescuers – Sorry
- Starbucks dropped the ball in New York (Charged Rescue Workers For Water To Save Lives)

The issue here isn't whether we should expect Starbucks, or any other company, to screen out employees who are capable of making what may be perceived as harsh, even callous decisions. Mistakes will happen. Rather, the question is why the company apparently failed, on several different occasions, to respond appropriately. At any one of the points when Midwood, or reporters, made contact with them,

Starbucks could have demonstrated the true nature of its corporate soul.

The customer service people might have been trained better to understand that they're in a unique intelligence gathering position, capable of detecting and reporting problems. The people who answered phones and letters for company executives, and the executives themselves, were in the same position. Company communicators can't respond or recommend the proper response to events that go undetected.

Only after the news stories ran did Starbucks apologize, hand-carrying a refund cheque from the company to Midwood. And Smith called Midwood's Rapisarda to apologize personally. But the damage had already been done.

The many stories covering the apology merely restated the story of the original sin. And its timing conveyed the impression of a cold, money-hungry company that had been caught pursuing the ruthless business policy that has resulted in the proliferation of Starbucks' coffee houses around the world.

Only a few stories mentioned that Starbucks gave coffee away to rescuers. Or that it spearheaded an effort that collected, by the time of the negative publicity, around $2 million for victims of the September 11 events. None of them mentioned the company's several corporate social responsibility programmes.

And therein lies the danger – when an essentially decent organization isn't equipped to respond to crisis and communicate in a way that its corporate creed dictates, it runs the very real risk of becoming known not for what it is, but for what it is not.

Adapted from an article by Phil Cogan, Bernstein Communications, Inc. (2001).

Government officials

Key members of this audience need to be kept appraised of the emergency situation, as determined appropriate by the Government Affairs Department.

Customers

Customers need to know that the company is concerned about crises that impact on its operating divisions and may affect production.

News media

We want the news media to know that the company is credible, concerned and effective at dealing with emergencies, and that we understand and meet the unique needs of each type of media.

Vendors, contractors and suppliers

We want these audiences to know that we operate state-of-the-art safety monitoring and control systems in our factories and laboratories and that the company takes quick, effective steps to protect the health and welfare of its employees, vendors and suppliers.

How people receive information during a crisis

In order to determine the most effective ways of communicating during an emergency, it's important to consider how each key audience potentially can receive information. The following sources of information for each audience must be considered in communication strategy for each emergency:

Employees

- direct knowledge of the event
- other employees
- people external to the company
- intercom and phone systems
- managers
- electronic mail
- news media reports
- fire department/police/hospital spokesperson

Community residents

- direct knowledge of the event

- neighbours
- news media reports

Top management

- company security, communication/PR manager
- personal call(s) from concerned employee(s)
- news media reports

Government officials

- government affairs
- news media reports
- concerned or scared citizen(s)

News media

- public relations representative
- police and/or fire scanners
- other news media
- fire department/police/hospital spokesperson
- eyewitnesses, including employees
- bystanders with knowledge or hearsay
- first-hand view of the situation
- outside 'experts'

Vendors, contractors and suppliers

- direct knowledge of the event
- employees and/or management
- news media reports

Evaluating probabilities

We believe that every organization should create for itself a 'crisis inventory' in which it identifies the sort of problems it may face in the normal course of business. Furthermore, in reviewing this list, it should assign a relative probability to each of the incidents, based upon the following scale:

High   Very probable; has occurred in past; has happened to other companies in the community or in your industry.

Medium   Less probable; no past occurrences within your company; has rarely occurred in community or industry.

Low   Very remote possibility; once-in-a-lifetime occurrence,
           yet within the realm of possibility.
None   Virtually impossible.
N/A   Not applicable to your business.

Note: this checklist is designed to encourage you to think about the very type of incidents that most companies prefer **not** to think about. Experience has shown that an 'it can only happen to the other guy' attitude is one of the major obstacles to crisis prevention and effective crisis management. You may also be surprised by your own probability assessments. Most likely, they will be higher than you anticipated (that's precisely the goal of the exercise – to increase awareness of the need for crisis planning).

  If all categories are applicable to your organization, you are encouraged to select one crisis from each category and think through the various guidelines offered in this book with those particular crises in mind.

Creating a 'Crisis Inventory'

It is suggested that the crisis management team work through the following inventory:

TYPES OF CRISIS   PROBABILITY

	HIGH	MEDIUM	LOW	NONE	N/A
Natural disasters					

e.g. flooding, earth tremor, wind, bushfire

	HIGH	MEDIUM	LOW	NONE	N/A
Plant operations					

e.g. fire, explosion, power failure, industrial accident, computer failure, industrial terrorism

Environmental accidents

_____

_____

_____

_____

e.g. release of toxic chemicals into air, rivers, sea, water table

Environmental liabilities (potential crises)

_____

_____

_____

_____

e.g. community exposed to pollution, hazardous waste, noise, community protest

Employee safety & health

_____

_____

_____

_____

e.g. exposure to chemicals, gases, waste, faulty equipment, personal injury suits

Customer relations

_____

_____

_____

_____

e.g. subject to product liability, recall, tampering, consumer boycott, service complaints and any negative rumours and publicity

Social controversies

_____

_____

_____

_____

e.g. regional, national, international, AIDS, action of parent company, business profile

Labour relations

_____

_____

_____

_____

e.g. negotiations, unfair labour practice, strikes, affirmative action, racism, equity, privatization, training

Management issues

_____

_____

_____

_____

e.g. retrenchment, downsizing/layoffs, competition, lawsuits, executive kidnapping, price fixing, trademark, patent or copyright infringement

Employee/management misconduct

_____

_____

_____

_____

e.g. bribery, corruption, kickbacks, industrial espionage, nepotism

Government affairs

_____

_____

_____

_____

e.g. legislation, lobbying

Other

_____

_____

_____

_____

Focusing on key scenarios

Now that you have assessed the probability of occurrence, the next step is to add the dimension of impact, or relative seriousness, for each event identified. You can use the chart below for this purpose to plot the five or six most probable crises which you believe your crisis communication plan must focus upon.

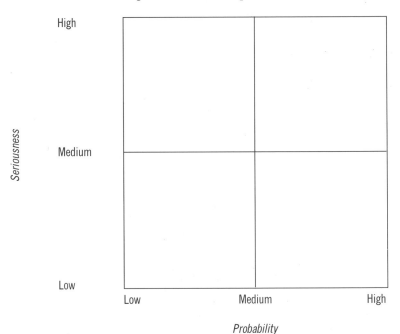

Figure 6
*Chart for plotting
probable crises*

You may wish to have various sections or departments of your organization fill out a comparable chart and then convert this information into a master chart for your company.

We believe that the most valuable benefit of this exercise is to use the list as the foundation for developing crisis prevention programmes. Many excellent books and articles which have been written on the subject of crisis prevention detail how to detect early warning signs – and implement preventive initiatives. Needless to say, the more each member of the crisis communication team understands about the total crisis management process, the better prepared your company will be to plan for and deal with probable crises.

In the next chapter we will deal specifically with crisis management preparedness and the need to review the existing management culture of an organization.

Sources

Barton, L (1993) *Crisis in organizations*, Cincinnati: SW Publishing.

Bernstein, J (2002) Bernstein Communications [Online] Available: www.bernsteincom.com

Fearn-Banks, K (1996) *Crisis Communications: A Casebook Approach*, New Jersey: Lawrence Erlbaum.

Grunig, JE & Repper, FC (1992) 'Strategic management, public and issues' in JE Grunig (ed) *Excellence in public relations and communication management*, pp. 117–157, Hillside, NJ: Lawrence Erlbaum.

Skinner, JC, Von Essen, LM & Mersham, GM (2001) *Handbook of Public Relations*, sixth edition, Cape Town: OUP.

4

Crisis management preparedness

Crisis development cycles

A crisis goes through various stages. Fink (1984) and others identify four. The first stage is the warning or prodromal stage where
a number of indications are given that something is problematic.
At the second stage, the acute crisis stage, some damage has
already occurred, but the situation is still controllable. By stage
three, the chronic crisis stage, Fink says 'all hell has broken loose'
characterized by extreme media attention, lawsuits, government
scrutiny and erosion of markets and customers. This stage can
linger indefinitely. Stage four, the crisis resolution stage, is a
return to normality. A well-organized and comprehensive crisis
management plan which covers all four stages is essential if a crisis is to be controlled and managed. An integral part of this plan is
the crisis communication plan.

Research by Venter (1996) on the life cycles of crises reviews
more than fifty major crises resulting from industrial and financial disasters. He identifies four areas, namely, operational risk,
control opportunities, chaos and repair. These life cycles are
depicted in Figure 7 on the next page. The vertical axis of the grid
represents an increasing level of intensity, while the horizontal
axis represents time elapsed.

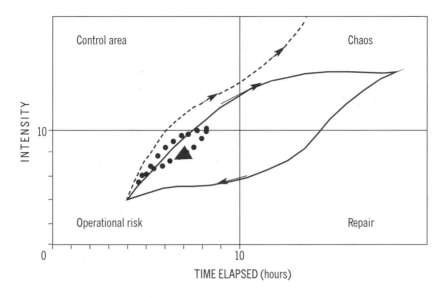

Figure 7
Life cycle of a crisis

Source: Lubbe and Puth (1994: 214)

The figure offers three possible variations of crisis development cycles, i.e. normal development cycle, an uncontrolled development cycle and the ideally controlled cycle.

In the normal cycle, a crisis occurs and moves briefly through a control phase; after which it enters the chaos area where it is arrested and forced into the repair phase and eventually restored to its position in the operational risk area.

The uncontrolled cycle speaks for itself. Its self-destructive curve soon escapes any form of control, and results in an organization losing all credibility and the ultimate shutdown of the business.

The ideally controlled crisis should be what every crisis plan strives for. In a sense, it is a contradiction in terms since the full impact of the potential 'crisis' is arrested at a point where little damage is done.

Snapshot: The need for an adequate emergency response plan

Some ten years ago, the local fire department servicing a municipal area to the north of Gauteng responded to a pre-dawn call. Two fire tenders arrived at the site of a chemical distribution warehouse which was thoroughly ablaze with part of the roof already collapsed. The alarm had been raised by security personnel at neighbouring business premises.

With the assistance of additional fire tenders from other fire departments, the fire was brought under control within two and a half hours by firefighters wearing breathing apparatus.

In excess of seventy different chemical products – principally veterinary and agricultural formulations – were stored in the warehouse, yet the company's management were unable to provide adequate information to emergency services personnel about the substances involved. Much of the information received was not accurate, leaving the emergency services personnel in the dark about the hazards they were dealing with.

Despite wearing breathing apparatus, twenty firefighters had to be admitted to hospital for observation. Chemical experts were commissioned to perform hourly monitoring while the contaminated fire debris was removed from the site.

No fire detection or protection equipment had been installed in the warehouse, although there were a number of fire hydrants around the building.

Despite radio and television broadcasts and newspaper reports there were numerous complaints from the public about the inadequacy of the information provided.

Debriefing by the emergency services after the event identified the following learning points.

- pre-planning by management must be a priority for sites such as this
- independent expert advice should be made available, if this becomes necessary

- testing of air, water and soil must be carried out during or as soon as possible after the fire to determine contamination levels
- sufficient personnel, equipment and the necessary training are high-level priorities
- there must be ready access to proper decontamination equipment and applicable instructions
- community evacuation plans must be formulated, disseminated and readily available
- electronic and print media releases alone are inadequate for communicating with the public during a disaster
- national building regulations do not provide adequate construction and safety standards for buildings of this nature and function.

Significantly, each of these points would have been identified in a comprehensive emergency response plan, had one existed.

Information supplied by Honi Brian, Cormark Communication, cormark@ircradio.co.za

Assessing a company's preparedness

The ability of a company to prepare for and cope with a crisis effectively will be shaped largely by the existing management culture of the organization. The fundamental strengths and weaknesses of a given company's management approach are greatly amplified in times of crises.

If a company has, for instance, not established close community ties, or if it has had a history of maintaining a 'low profile' on environmental matters, much more radical attitudinal changes on the part of management will be required in developing an effective crisis plan. Similarly, if a company has a highly decentralized structure, with each local site responsible for its own community and/or media relations, the concept of central and/or corporate-based control may need to be adjusted accordingly if the crisis plan requires centralized coordination.

Other organizational cultures to be taken into consideration in the planning process include the fundamental decision-making process in an organization, the role of internal functions (such as legal, public relations, government affairs, and insurance) in decision making, the degree of employee participation within a company, labour-management relations and ownership structure.

The first step in assessing the preparedness of an organization to develop and implement a successful crisis management programme is to analyze current management policies, organizational structure, communication systems and the attitudes of the management team – and evaluate whether or not these are compatible with the specific policies and procedures of the company's crisis response and communication plans.

Cultural obstacles to effective crisis planning

Certain attitudinal traits can prove to be obstacles in both the planning and crisis management processes, and can turn a problem into a public relations disaster. Companies are encouraged to analyze their own cultures in the light of the following traits and, where necessary, determine what policy changes may be required in order to gain company acceptance of the underlying philosophies of the crisis communication plan.

Isolationist mentality

Characteristics

The company views itself as an 'island', with little or no contact with the community. Managers are not active in the community. There is an absence of external research data concerning the attitudes of the community, community leaders, citizens' groups, or even employees, towards the company. Media relations are poor or non-existent. Potential problems and liabilities are kept secret and are regarded as 'proprietary'. Management believes it is solely 'in the business of doing business', and if some of its activities are 'dirty' as a result, 'that's the way everybody does business in this sector'. Management believes that what they do is none of the business of the community or the media and that it is within their power to keep it this way. Frequently, management believes that the public or community would not understand certain 'technical' information in any case.

Possible cultural adjustments and actions

Top-level reconsideration of the business mission and how each of the company's stakeholders might be impacted by a crisis is required. Management-level reorientation and education are the first steps. The following steps might be considered:

- Establish a management committee for 'social responsibility'; hold periodic meetings and workshops.
- Develop a mission statement that articulates the mission and values of the company. This could be shared with all managers and employees so that it would serve as a basis for fundamental decision-making and a policy foundation for community and crisis programmes.
- Establish a programme to monitor attitudes of various company constituencies – employees, the community, consumers, shareholders, etc. Such a programme might include community surveys or focus groups, external public relations, communication with pressure groups, etc.
- Invite community leaders and members of the local media to regular individual background briefings on company's operations.
- Appoint a manager to function as the resident 'devil's advocate' or develop relationships with third-party observers who could function as the company's 'social conscience'.
- Encourage managers to run the company 'as if the whole world were watching'.

Snapshot: What Drucker says

Gurus, sages and all manner of thinkers and theorists lay claim to priceless insights into the mysteries of business management and the production of wealth. Armed with case studies of their headlined successes they all claim to be the best communicators and motivators in the business. But few, if any, would dare to compare themselves to the management consultant Peter Drucker who invented business concepts like privatization, outsourcing, management theory, knowledge workers and the global economy.

Drucker says 'most people will make their living either working in or with organizations'. Drucker believes that an essential element of management should be a focus on an understanding of the relationship of the individual to the larger organization and to the world that surrounds them. 'What is lacking is the relation of themselves to the universe of knowledge,' he says.

'Economics, law, some history some structure of society and world economy, psychology – these are core disciplines. Without such social skills the executive is simply not effective in an organization. There is a great deal of emphasis on leadership today and I am all for it. But what about "followership?" Most people, even the so-called leaders, spend some time being followers and equal partners. What about learning that in an organization one doesn't begin by asking: "What do I want?" but by asking: '"What is needed?"'

Adapted from 'Time to rethink the curriculum', *Business Report*, 24 January

Reactive mentality

Characteristics

Business problems and crises are dealt with on a purely reactionary basis. No programme is in place to assess existing vulnerabilities or anticipate potential problem areas. Little attention is paid to trends in industry or society. The general attitude is that crises happen only to the 'other guy'.

Adjustments/actions

- Conduct an audit of 'public relations vulnerabilities', including areas such as major employee safety issues, potential for a product recall, transportation of toxic chemicals.
- Establish issues-monitoring programmes; track industry trends, news reports about occurrences at similar companies, and community or market concerns. Obtain background information on key issues from trade associations or your local chamber of commerce; subscribe to industry-sponsored public opinion research.

- Develop company responses to potential problem areas.
- Identify management personnel best qualified to deal with each specific issue; identify external resource experts.
- Make issues monitoring an ongoing management function, with regular reports and management briefings.
- Develop a crisis communication plan.
- Train and update management on the mechanics of implementing the plan.

'Them vs. us' mentality

Characteristics

Organizations that raise a concern about the company's social responsibility performance (in areas such as the environment, community support or employee health and safety) are viewed as 'the opposition'. Management's knee-jerk reaction is not to respond to their allegations or take them seriously, assuming that the public will also dismiss such groups as activists. Some company managers lump the media into the same category.

Adjustments/actions:

Broaden management's perception of the issue to take into consideration how it affects all involved constituencies. Include employee, environmental and community impact statements in controversial plans and proposals.

- Identify all local and national activist groups that have an interest in those aspects of your operation identified as 'public relations vulnerabilities'.
- Familiarize management with the goals and agendas of community and activist groups.
- Consider a programme of open dialogue with local groups to determine their viewpoint, explain the company's position, and possibly develop coordinated responses to problem areas.

'Don't tell the boss' mentality

Characteristics

Bad news gets filtered as it moves to the top of the organization. Local management tries to keep all problems hidden and localized. Past experiences have shown that management tends to 'shoot the messenger' in such situations. Potential problems are minimized and frequently 'polished' by overly optimistic, unrealistic assessments and forecasts.

This can have serious repercussions in times of crisis. There are many documented cases of 'early warning signs' of pending disasters that were kept from top management. When crises do occur, this attitude could also result in costly delays in bringing proper corporate and community resources to the problem.

Management must honestly assess whether it is inadvertently sending signals that inhibit the free flow of negative information. It must demonstrate to middle managers and supervisors that open and candid communication is part of responsible leadership.

Adjustments/actions:

- Develop a written statement on the company's communication policy; initiate other actions to elevate the importance of open communication at all management levels.
- Empower members of the crisis response and crisis communication teams as well as other key plant personnel (such as security) to communicate directly with top management when they believe the situation warrants it. In large organizations, this may mean bypassing existing reporting levels.
- Establish emergency preparedness review panels with regular briefings to top management.

'PR is not important' mentality

Characteristics

The public relations function is non-existent within the company or viewed as a relatively low-level 'service' operation. Public relations personnel are viewed as 'technicians' as opposed to management and strategists. Budgets and personnel are typically cut when business is down. The company's relations with employees, the community, and media are not planned or managed in a strategic manner. When problems occur, top management tends to turn to legal counsel first. Public relations is not highly regarded – or trusted – because it is not viewed as a quantifiable function. Issues management and planning for unknown and uncontrollable contingencies are viewed as ambiguous activities of little value to the company's bottom line.

In times of crises, such companies typically lack the professional public relations expertise to properly manage community or media relations. Management and its legal counsel are reluctant to bring in an outside public relations firm. When, and if, such counsel is brought in, the legal function typically oversees all public information activities.

Adjustments/actions

While not all companies can afford to have full-time public relations personnel on staff, it is prudent to develop a relationship with competent public relations professionals before a crisis strikes – individuals who have an established track record in the successful handling of corporate crises. Such professionals can be identified through your local branch of the Public Relations Institute of South Africa (PRISA), trade associations or the local chamber of commerce. Firms with established public relations operations are well prepared for most crisis contingencies. The following strategies could be implemented to strengthen the relationship between top management and the public relations function:

- Initiate joint management/public relations education programmes, including issues management workshops.
- Brief public relations personnel, on at least an annual basis, on the company's employee health and safety programmes, environmental compliance programmes, transportation and distribution systems, legal priorities (including the status of law suits), as well as general business and marketing plans.
- Review the respective roles of public relations and legal personnel with regard to the release of information to ensure the most productive and expedient method for communicating with the public in times of crisis.
- Insist that public relations professionals be notified immediately of potential, impending, or occurring crises.
- Agree that all public documents be developed jointly by public relations and legal professionals. (Sending documents such as news releases back and forth between management and the legal and public relations functions can impede the communication process during times of crises.)

Each of the above cultural conditions can have a significant impact upon a company's ability to plan for and implement an effective crisis communication response. It is therefore important that all companies have a thorough understanding of their specific management culture before attempting to construct a response mechanism.

Sources

Fink, S (1984) *Coping with Crisis*, New York: Nations Business.

Lubbe, BA & Puth, C (eds) (1994) *Public Relations in South Africa: A Management Reader*, Isando: Heinemann.

Regester, M & Larkin, J (1998) *Risk Issues and Crisis Management*, second edition, London: Kogan Page.

Venter, N (1996) 'Crisis communication' in Lubbe, BA & Puth, G, *Public Relations in South Africa: A Management Reader*, Chapter 13, Johannesburg: Heinemann.

5

Preparing the plan

Aims of emergency plans

The purpose of establishing a crisis and disaster response programme is to help ensure that a company is prepared to deal effectively with unexpected events – and thus minimize adverse consequences.

The goals of emergency response planning are to:

- prevent fatalities and injuries to employees and members of the public
- provide guidelines for decision making
- identify and clarify responsibilities
- ensure that valuable time is not lost in implementing or coordinating response efforts
- minimize production downtime and disruption of business

What to include

A well-developed crisis communication plan should clearly state company policies and procedures under emergency conditions and provide the specific information required to carry these out.

In the succeeding pages we take you through the following key elements of a crisis communication plan:

1 purpose of the plan; the company's philosophy and policies towards stakeholders
2 agreement by top management to the above and selection of members of the emergency response team

3 explanation of specific responsibilities for each team member
4 exchange of all communication contacts between members of the team
5 listing of all local emergency personnel, local officials, hazardous-materials directors, and other authorities
6 listing of key media personnel
7 physical description of on-site and/or off-site crisis control room and a list of required equipment
8 media response plan, including delineation of responsibilities, prepared press materials, location and requirements for press room, selection of primary spokesperson, etc.
9 description of crisis communication network, including telephone 'pyramid' system and other procedures
10 a training programme, including orientation with the plan manual, emergency response, and communication and/or media training
11 a testing programme, including periodic testing of communication alert system (internal and external), team assignments, and mock drills
12 inventory of potential crises and public relations vulnerabilities
13 guidelines for communication during a crisis
14 suggested communication activities for crisis follow-up

Statement of purpose

The purpose of an emergency response plan is to ensure that appropriate actions are taken by responsible personnel in a timely manner, in order to minimize the impact of accidents, natural disasters, extraordinary business events, and related activities upon human life, the environment, the community, company property, and normal business operations.

The purpose of a crisis communication plan is to ensure that all personnel in a position to contain and manage a given crisis are provided with the information they require for a swift and effective resolution of that crisis, and that all other affected people are given factual information about the crisis as quickly as possible.

Policy and philosophy

A crisis plan should be based upon the principles and policies that guide a corporation in its normal business operations. Below are several sample statements that are typically found in corporate emergency response and crisis communication plans:

- While the goal of the corporation is to generate value for its shareholders by meeting the needs of customers, it recognizes its responsibility to operate in a manner that does not jeopardize the safety and welfare of its employees or the general public, and does not have a negative impact upon the environment.
- In times of crisis or emergency, human life and the environment take precedence over company property and business operations.
- Each member of the corporation is required to uphold the laws of the land and act in an ethical manner.
- The corporation has an obligation to cooperate fully with local authorities on all matters that impact on the community.
- The integrity and reputation of the company, as both a community and industry citizen, must be upheld.
- The corporation is committed to full, candid and timely disclosure of facts that the community and/or media have a right to know (particularly information relating to public health, safety and the environment).

Many companies base their emergency response policies and procedures on the principles inherent in their company's mission statement or credo.

Setting up an emergency management team

A small team of senior executives should be identified to serve as your organization's crisis communication team. Ideally, the team will be led by the company CEO, with the firm's top public relations executive and legal counsel as his or her chief advisers. If your in-house PR executive does not have sufficient crisis communication expertise, he or she may choose to retain an agency or independent consultant. Other team members should be the heads of major company divisions, to include finance, personnel and operations.

Initial crisis-related news can be received at any level of a company. A security guard or cleaner may be the first to know there is a problem, or any staff member might receive an anonymous phone call. All members of an organization should know who should be notified, and how they can be reached. An emergency communication contact list should be devised and distributed to all company employees, telling them precisely what to do and who to call.

A typical emergency response team might include the following positions:

1 Crisis manager (local site)

functions as the highest ranking on-site manager responsible for general coordination, decision-making, and communication with top management (if appropriate) and local officials. If the crisis is company-wide, this individual should be a highly ranked corporate officer.

2 Assistant crisis manager

assists the crisis manager, and assumes management responsibility if the crisis manager is unavailable.

3 Second assistant manager

assumes responsibility if the assistant manager is unavailable.

4 Emergency personnel coordinator

serves as primary contact and co-ordinator of emergency personnel.

5 Crisis control room coordinator

takes responsibility for physical setup of internal communication headquarters.

6 Public relations coordinator

assumes overall responsibility for all communication with internal and external audience. Also advises the crisis manager on strategy, prepares press materials and statements, coordinates media contacts, and maintains contact with the corporate public relations department. He/she may assign special communication duties to other team members for the following functions:

- **Community relations** – including local elected officials, chamber of commerce, citizens' groups
- **Employee relations** – responsible for letters to employees, workplace meetings, communication with families
- **Union relations** – including communication with both union management and the rank and file
- **Government affairs** – including contacts with high-ranking government officials, state and provincial legislators
- **Media relations** – could include several individuals assigned to specific media beats (local news, financial, trade)
- **Security** – for access to victims, access to the site by rescue workers, media, etc.
- **Clerical/administrative assistance** – typing and mailings, phone contacts, courier service
- **Marketing/customer relations** – customer inquiries, statements to sales organizations, dealers/distributors and customers
- **Investors/financial relations** (could be a local or corporate staff position) – inquiries and other communication with shareholders and the financial press
- **Press centre coordination** – setting up, equipping, supplying and maintaining the press room
- **Corporate relations** – if the site or plant is part of a larger corporate organization, ongoing communication with corporate management may be necessary.

7 Media spokesperson(s)

Within each team, there should be individuals who are the only ones authorized to speak for the company in times of crisis. Communication skills are one of the primary criteria in choosing spokespersons. The CEO should be one of those spokespersons, but not necessarily the primary spokesperson. In extreme crises, however, it is absolutely necessary for the CEO to speak out. Usually in a crisis there is too much communication for one spokesperson to handle alone. However, one individual should be the primary person who speaks to the media. The spokespersons should have direct access to management and technical advisers, and should guide other spokespersons as necessary. Your spokespersons should have professional training in how to speak to the media.

8 Technical adviser(s)

serve as primary information 'resources' to the media spokesperson and public relations personnel. They can also serve as 'expert' media spokespersons, backing up the primary spokesperson. They should be the most knowledgeable individuals for the given situation, familiar with plant processes/systems and company operations. Areas that may need to be considered include:

- manufacturing operations
- chemicals/materials
- environmental compliance
- utilities
- industrial safety
- employee health/toxicology
- marketing
- security
- transportation
- information technologies

Note: It may also be prudent to identify third-party advisers (independent experts) who can provide objective information and opinions to the media.

9 Legal adviser

provides ongoing counsel to crisis manager, corporate managers and public relations personnel.

10 Evacuation coordinator(s)

coordinates evacuation plan with local authorities. May also have other team members responsible for transportation, communication, temporary housing for displaced employees, food, water, clothing, etc.

11 Plan editor and coordinator

is responsible for maintaining, updating, and distributing written plan. All changes in response team duties, phone numbers, and addresses should be forwarded to this individual as they occur.

12   Corporate management contacts (if appropriate)

If the crisis is of a company-wide nature, various ranking corporate officers may be required to take an active role in both crisis containment and communication.

13   External public relations counsel

Whether or not you currently utilize a public relations agency or consultant, it is advisable to identify in advance qualified and experienced external public relations professionals in each of the regions where you have facilities, should you require professional public relations counsel and media assistance.

Emergency response team

Last updated:

The following individuals will comprise the emergency response management team for this site/company:

Crisis manager _____

Assistant _____

Second assistant _____

Emergency personnel coordinator _____

Back-up _____

Crisis control room coordinator _____

Back-up _____

Public relations coordinator _____

Back-up _____

Media spokesperson _____

Back-up _____

Technical advisers _____

_____

Legal adviser _____

Back-up _____

Evacuation coordinator _____

Back-up _____

Plan editor/coordinator _____

Optional

**Corporate management
contacts** _____

_____

_____

_____

**External public relations
counsel or agency** _____

_____

_____

Other personnel _____

_____

_____

Involvement of top management

One of the most critical decisions that must be made in the early
moments of a crisis is whether or not to involve top management
and, if so, the role that top management should play in the resolution of the crisis.

Experience with several recent industrial crises has shown that non-involvement by the top ranking officer of the corporation can waste time, impair decision-making, and jeopardize the company's credibility.

If the crisis warrants top management involvement, it is important that the president or CEO takes an active and visible role in the resolution of the crisis immediately.

Some companies have established a list of clearly defined emergencies that require top management involvement (e.g. multiple injuries, transport accidents involving the release of toxic chemicals into environment) while others have adopted policies requiring that all 'non-routine' incidents be automatically reported to a central corporate office for 'assessment'. What works best will depend upon a variety of factors. It is important that this policy be defined in advance and included in the crisis plan.

Establishing a crisis communication network

In order that all 'need-to-know' personnel are alerted to the crisis as soon as possible and that communication throughout the crisis remains open and consistent, it is necessary to set up the following communication provisions:

A networked system includes landline, cellular phone and e-mail networks, with the emphasis on the sequence and flow of contact. Names of all emergency team personnel should be provided to all team members. Each member should be given responsibility for calling two or three additional team members.

Here is how one such network might work:

1 First person aware of crisis contacts:
   - emergency personnel coordinator
   - crisis manager
2 Emergency personnel coordinator contacts:
   - emergency personnel only
3 Crisis manager contacts:
   - assistant crisis manager(s)
   - crisis control room coordinator
   - evacuation coordinator
4 Assistant crisis manager contacts:

- public relations coordinator
- chief technical advisers

5 Public relations coordinator contacts:
- corporate public relations contact (if appropriate)
- media spokesperson
- press relations coordinator

6 Additional members of public relations team contact other team members

7 Chief technical advisers contact:
- other technical advisers

The database of names and phone numbers of the entire team, other key personnel in the organization and local emergency personnel should be stored in a number of locations, for example, on laptop and PDA electronic address books, in cellular phones, and on paper.

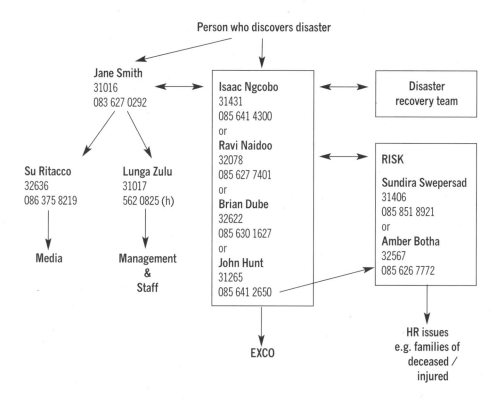

Figure 8 *Example of a crisis communication network*
Source: BoE (Board of Executors)

Setting up a crisis control room

Emergency Operations Centre

An Emergency Operations Centre (EOC) is a centralized location from which emergency response during a very severe crisis is coordinated and directed. A crisis that would force a site EOC to be engaged would be an incident such as an earthquake or an act of terrorism that results in multiple crises.

If a site EOC is engaged, all crisis communication will be directed from that point. For adequate communication response, two communicators should be assigned to the centre. One communicator would serve as the Communication Manager who is responsible for coordinating external and internal communication which may include the use of a runner system, in the event that all electronic means of communication are down. The other communicator would serve as the Information Officer, responsible for documenting the sequence of events in support of communication and then generating any statements for release either internally or externally.

Each site EOC should be equipped with the following:

- networked computers with Internet and e-mail facilities
- maps
- photocopiers
- telephone lines
- fax machine(s)
- flipcharts, whiteboard, stationery and pens
- laptop computers
- printers
- cellular phone(s)
- radio and television receivers, videotape player
- division crisis communication plan
- microphone or bull horn
- name tags
- refreshments

Establishing a media centre

A predesignated, convenient location for greeting and briefing the media on an ongoing basis during the crisis is critical. Generally,

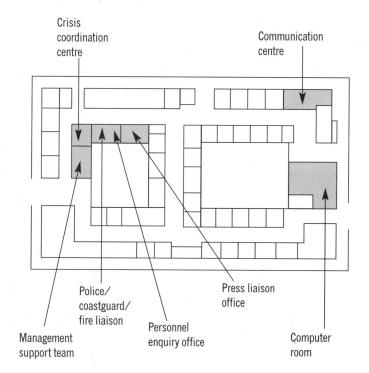

Crisis
coordination
centre

Communication
centre

Police/
coastguard/
fire liaison

Press liaison
office

Management
support team

Personnel
enquiry office

Computer
room

Figure 9
Typical crisis
coordination facilities
Source: Shell Public Affairs
in Practice

cost and practical considerations will dictate that the Emergency
Operations Centre and Media Centre will be at the same location.
However, care should be taken that they can be physically separated and access controlled.

This centre could also serve as the command centre for the
company's public relations personnel. Ideally, the media centre
should have at least three rooms or offices:

- a conference room with seating for at least 20 people for conducting press briefings
- an office (or offices) equipped with the communication
  requirements described above for the EOC
- a separate room for private interviews.

Identifying 'need-to-know' audiences

During and after a crisis, there are a number of people – internal
and external – who must be communicated with in a timely manner. A chemical spill, for example, would require immediate
notification of community authorities, while a product recall
would warrant immediate communication with distributors and

retailers. It is essential that updated names and phone numbers be kept in the crisis response plan and that each team member understands his or her specific communication responsibilities.

The following is a representative listing of individuals who may need to know about a site emergency: (The listing does not suggest order of contact.)

	Name	Phone
1 Local emergency personnel		
Police		
Paramedics		
Hospitals		
Fire		
Local health department		
Local environment agencies		
Electricity supplier		
Water supplier		
Sewage supplier		
Other		
2 Government officials		
Mayor		
City/town officials		
Provincial officials		
Government representatives		
3 On site management and employees		
Labour relations		
Personnel		
Operations		
Engineering		
Maintenance		
Marketing/sales		
Secretarial/administrative		

	Name	Phone

4 Union officials

_____ _____ _____
_____ _____ _____
_____ _____ _____
_____ _____ _____

5 Corporate management

	Name	Phone
CEO		
Human resources		
Public relations		
Financial relations		
Government affairs		
Engineering		
Legal		
Insurance		

6 Marketing/sales/key customers

_____ _____ _____
_____ _____ _____
_____ _____ _____
_____ _____ _____

7 Key vendors/suppliers

_____ _____ _____
_____ _____ _____
_____ _____ _____
_____ _____ _____
_____ _____ _____

8 Other community contacts

	Name	Phone
Chamber of Commerce		
Economic development groups		

9 Citizen groups

_____ _____ _____
_____ _____ _____
_____ _____ _____

Monitoring the attitudes of your publics

Ideally a crisis communication plan should include some means of gauging the attitudes and opinions of groups affected by a crisis: employees, shareholders, the community, consumers, etc.

Many companies routinely conduct employee and community surveys to get a current understanding of how these important constituencies perceive them. Such surveys also help to point out potential concerns – before they take on crisis proportions.

There are a variety of circumstances where documentation of employee or community support can prove highly valuable in times of crisis.

The ability to sample community opinions quickly can also prove useful in creating pre- and post-event benchmarks of public attitudes.

In certain instances, sampling of public opinion may be necessary to determine if the affected public perceives the incident as a crisis and whether a crisis response is required at all.

Supplier 'swat' team

Emergency situations frequently demand overnight and/or weekend production support from public relations and advertising agencies, typesetters, printers, photographers, artists, media buyers, direct mail houses and related communication specialists. Not all vendors are equipped – or willing – to work on such a basis. Much time can be lost trying to track down the home phone numbers of people who can be counted on in an emergency. Therefore it is a good idea to set up this support network in advance.

	COMPANY	CONTACT	PHONE	HOME
1 PR agency				
2 Advertising agency				
3 Typesetter				
4 Printer				
5 Layout artist				
6 Media buyer				
7 Direct mail				
8 Media wire service				
9 Web master (if outsourced)				

10   _____ _____ _____ _____
11   _____ _____ _____ _____
12   _____ _____ _____ _____

Developing key messages in advance

Because you have already identified what type of information audiences are looking for in the audit and planning stages, it is possible to devise basic messages in advance of any potential disaster. Although such messages will need refinement should a crisis occur, much time can be saved by proactive preparation. Delay-related damage caused by lack of planning can easily double or triple the time and cost of damage control. Delay can also result in irreparable harm. At the same time, it takes only a one-time plan, with minor updating, to serve as a template and operating basis for all future crises.

Writing and reviewing the plan

It is recommended that an editor be assigned to the writing and editing of the plan. Drafts should be reviewed by all site as well as key corporate managers.

     The editor should update all names, phone numbers and addresses immediately upon receipt of changes and issue change notices or, preferably, reprinted insets. The editor should also maintain a master distribution list. We recommend that the plan should be reviewed thoroughly at least once a year.

Sources

Cutlip, SM, Center AH, & Broom, GM (2000) *Effective Public Relations*, New Jersey: Prentice Hall.

Shell, *Public Affairs in Practice*, section on Crisis Management, London: Shell.

Skinner, JC, Von Essen, LM & Mersham, GM (2001) *Handbook of Public Relations*, sixth edition, Cape Town: OUP.

Skutski and Associates Inc. & Communications (1990) *Crisis Communications Planning Guide*, a step-by-step workbook for preventing public relations disasters, New York: Skutski.

6

Training and testing

By now, it will be clear to the reader that the crisis and disaster management plans are quite complex and require excellent coordination of human and material resources. It will also be evident that communication is the 'glue' that holds the process together. This said, it is also obvious that such complexity and rigid coordination requirements will benefit from training and testing. Training and testing are important also from a psychological point of view because individuals involved will have an enhanced feeling of being able to cope, of being able to implement initial actions which will relieve stress and also reduce the need for unseen problem solving.

Training and testing

1 Emergency response management team

Each member of the emergency response team should receive training in the following areas so that they:
- thoroughly understand company policies
- are familiar with all aspects of the crisis response plan
- undertake a review of crisis scenarios and related public relations vulnerabilities
- review the communication network responsibilities
- review local and provincial emergency response capabilities and personnel

- are well versed in site processes, hazardous materials, utilities, etc.
- are apprised of public and legislative hearing processes
- are familiar with 'right-to-know' and related laws that affect the site
- participate in mock disaster drills
- participate in an annual team orientation and review meeting

Selected team members should receive additional training or knowledge about:
- emergency first aid, including CPR
- presentation skills training
- media training
- legislative policy review
- corporate legal policy review

Many of these topics could be addressed in an internal crisis prevention and response workshop, which could also serve to bring your team together as well as to underscore management's commitment to crisis planning. Periodic issues-management updates, instructional videotapes, workshops, a crisis response team newsletter, and related training techniques could also be utilized to keep your crisis team in a state of readiness.

2 Communication network

Each member of the team must be made aware of telephone contact responsibilities. It is extremely important that all team members recognize the importance of informing their support staff, such as personal assistants and secretaries, as well as a member of their household, of their role in the plan and the need to refer emergency calls to a designated back-up person, should the team member be unavailable.

3 Media training

Management personnel, the public relations staff and the designated spokesperson will most likely be responsible for the media relations during a crisis. However, it is recommended that all members of the emergency response team be provided with media training. Familiarizing all members with the workings of the media will help streamline the flow of information during a crisis. This training will also stress the importance of complete and timely disclosure of available facts in a manner that is clear, concise, and understandable to the media and the public.

A pre-designated primary spokesperson should be assigned to the emergency response team, but the key spokesperson for a particular crisis may be determined by circumstances. For instance, this spokesperson may be a technical expert or a personnel manager.

The media training process will help to clarify roles and responsibilities and provide a means for testing the plan under simulated conditions.

Testing

Once the plan is written and approved, it is recommended that each facet be tested.

Telephone/communication system

A simple call-back test of the telephone communication network (pyramid system) should be conducted during both business and non-business hours. The network should be activated with a single phone call. Each team member should make his or her pre-assigned calls, and then call the test coordinator to verify completing the calls or to report any difficulties encountered.

Emergency response assignments

Team members' knowledge of their specific responsibilities can be tested by telephone inquiries or written tests, or in group seminars.

Mock drills

The most effective way of testing the plan is with mock disaster drills.

A single simulated crisis exercise can be conducted in a workshop setting and be based upon previously identified crisis scenarios. Some companies have designed more elaborate exercises, involving numerous role-playing exercises – even hiring 'actors' to play the roles of the media and elected officials. Others stage full-blown mock drills in cooperation with local government, hospitals and emergency personnel officials. Assistance in simulated crisis training may be available through local government agencies and emergency personnel organizations. Public relations agencies experienced in crisis management also conduct simulated crisis exercises.

Every organization should set aside one or two days a year to prepare itself for an emergency. Here are suggestions on how an introductory course could be organized.

The best approach is to start with a general analysis called 'Anatomy of Crisis'. Here participants analyze various crises, and why the outcomes of similar crises can vary extensively.

After the critical 'Anatomy' session, participants are provided with guidelines for preparing for and handling a crisis. They are then broken into groups, each group receiving a different scenario to work on over an extended lunch period. It is important to provide each delegate with clear written instructions on what his or her group is expected to achieve and the need to appoint a group leader and spokespersons.

After lunch, the team leaders or spokespeople report back after presenting their scenarios. These scenarios are carefully devised to foreground the dilemmas that a crisis team faces in real life. This ensures that by mid-afternoon every delegate not only understands the general principles, but he or she also realizes that there are no simple answers.

Once the groups have selected a winner, it is time to move on to guidelines for handling the media, followed by a spokesperson from each group giving a no-holds-barred TV interview on the subject of the group's scenario in front of other delegates.

Conducted in a spirit of enjoyment and humour, the end result should be that relevant managers rethink their crisis programme in the light of changed circumstances.

Extending training over two days, however, allows for further discussion on unfolding scenarios and more specific and intensive media training. In the latter case, delegates can be put through hostile interviews for both print and electronic media and taught how to deliver the key messages designed to position the company positively. Clear direction can also be given in understanding the media agenda and methods and in using these to the best possible advantage. To give the exercises more credibility and relevance, the teams could be invited to a television studio or independent operator to be interviewed by a media personality.

Manual use and distribution

The written plan should be distributed to all members of the emergency response team and all managers and supervisors of emergency response team personnel. Copies should be placed in all key locations. All supervisors should be made aware of the existence of the manual in the event that they need to contact emergency response team members.

Distribution and updating should be assigned to the plan editor.

In Chapter 8 we discuss details of an actual Disaster Management Plan which has been drawn up for one of the major industrial complexes in South Africa.

SNAPSHOT: An industry-specific initiative

The prevalence of incidents involving chemical products, processes and plants during the latter half of the 20th century led to increasingly negative sentiments towards the industry. There was growing realization that severe ecological damage invariably accompanied the pollution that followed chemical spills and gas emissions. This threat to the environmental integrity of the planet was compounded by the potential for harmful effects on humans, particularly those close to chemical plants. The chemical industry became the target of sustained criticism, of which a fair proportion was probably justified in the light of some of the practices that came to light.

Responsible care

Realizing that it had become vitally important to restore the industry's credibility, the Canadian Chemical Producers Association (CCPA) embarked during the mid-1980s on an initiative that was intended both to improve their environmental health and safety performance and to communicate this fact to the public. In much the same way as the United Nations Environment Program launched APELL, this came on the heels of a series of major chemical disasters around the world.

In a proactive move, the association advocated sharing performance and emergency plans with the public to alert them to potential hazards, thereby minimizing the impact on human health and safety and on the environment.

This initiative, known as Responsible Care, has since been adopted by the chemical industry internationally. It has been described as an ethic and a way of life, requiring total commitment and a culture change to continuous improvement of performance in health, safety and the environment. It represents, in addition, a commitment to the principles of responsible management and stewardship of chemical products and services 'from the cradle to the grave' – from the research and development of a new product through to its final end use and disposal. It requires also open and honest communication with the public and the sharing of information with surrounding communities. Commitment is signified by a company's Chief Executive Officer publicly signing the guiding principles.

The South African Chemical and Allied Industries' Association (CAIA) adopted Responsible Care as one of its main programmes in 1994.

CAIA represents the majority of chemical companies that account for some 90 per cent of the chemicals manufactured in South Africa and boasts a long list of Responsible Care signatories from the storage, transportation, trading, manufacturing and distribution sectors who have voluntarily undertaken to be proactive and respond to public concerns about the manufacture, transport, use and disposal of chemical products and packaging.

The CAIA programme is based on guidelines provided by the International Council of Chemical Associations and on models used by sister organizations in Australia, Britain, Canada and the USA. It embodies six elements:

- health and safety of persons
- storage and distribution of chemicals
- transportation of chemicals
- waste management and pollution control
- community awareness and emergency response
- product stewardship

These combine to form a holistic approach, highlighting the principle that proactive planning for environmental health and safety is essential to ensure sustainable development. CAIA provides member companies with guidelines for implementation and biennial self-assessment questionnaires to monitor progress.

Quantitative indicators of performance have been developed to publicly demonstrate performance, and revisions are made as necessary to keep the South African industry in line with international benchmarks.

As in the case of the APELL programme, the role of communication is emphasized throughout, both internally with employees and externally with all stakeholder groups. Comprehensive guidelines cover communication channels, tools and techniques that can be used proactively to create awareness as well as to facilitate the management of information transmission during a crisis. Finally, though, CAIA cautions that good communication cannot compensate for poor health, safety and environmental performance and that demonstrably improved performance will remain the cornerstone on which the effectiveness of the Responsible Care initiative rests.

Community Awareness and Emergency Response (CAER)

In common with the APELL process, the CAIA model considers CAER programmes to be of primary importance and, in fact, obligatory on the part of facilities that manufacture, process, use, distribute or store hazardous chemicals.

Signatory companies are expected to recognize the need and right of the public to know the risks associated with operations and chemical products manufactured in or transported through its communities, and to undertake to maintain ongoing open and honest dialogue with employees and the community. Member companies are also required to be sensitive to the public's concerns and questions and to let them have information about the company's safety, health and environment programme activities and performance.

The value of community outreach

CAER programmes are by definition exercises in proactive communication intended to :

- promote community knowledge through outreach

programmes and consultative mechanisms for employees and the community

- ensure emergency preparedness so as to protect employees and communities in the event of an incident
- create mutual understanding by stakeholders of rights, responsibilities, concerns, needs, resources and mutual benefits
- facilitate cooperative sharing of resources and facilities, through joint plans and actions that serve not only to protect the community, employees and the company, but also to safeguard the environment
- increase the confidence of the community and employees through knowledge of hazards and of the necessary precautionary safeguards that are in place
- promote a positive company and industry image within the community

Proactive interaction with the community has a significant benefit: it is suggested that, for the industry to counter perceptions that it is an uncaring polluter of the environment, characterized by lack of openness and frankness about operations, it needs to start at local level. Gaining the confidence and trust of the local community will do much to improve the reputations of the company and the industry; and positive community support can act as a buffer to alarmist pressure tactics or reporting.

Achieving the above objectives will mean establishing active channels of communication with the public and with employees, and setting objectives and formulating plans for community consultation and an operational community emergency response plan.

There have been notable successes. One that comes to mind is the positive impact that Dow, a global chemical manufacturing group which has recently repositioned itself as a science and technology company, appears to be making with its community outreach programmes, in the process setting new standards for companies worldwide. In its 2000 Public Report which details the company's progress on public policy issues, Dow indicates that it aims for the results of its periodic Community Perception Surveys to show that at least 80 per cent of residents and leaders in each location where Dow has a significant presence agree that Dow is a good neighbour and a valuable member of the community. These surveys cover topics ranging from economic impact on the community to public health and safety, contributions spending and hiring practices.

These are valuable measuring instruments that also identify areas of economic, environmental and social responsibility that require improvement. In collaboration with its Community Advisory Panel each location is responsible for meeting the needs identified through the survey results, by developing and implementing its own action plans, through community dialogue and other forms of communication. The results of the 2000 survey were impressive with six out of seven global locations surveyed at that stage returning favourability rating scores between 71 per cent and 85 per cent.

By contrast, there have been many quoted instances of retaliation against companies who were considered by their stakeholders to have been remiss about communicating necessary information before the occurrence of an incident. Within the chemical industry context, the Bhopal disaster is a case in point.

Stakeholders who felt that they had been 'buffered out of existence', i.e. not told the full truth prior to the crisis, retaliated strongly against the corporation, as demonstrated by the lawsuits against Union Carbide by its shareholders for 'basic misrepresentation of the dangers involved in operating the plant'.

The effects of inadequate communication go beyond legal action, however, with potentially serious consequences in the form of loss of life and property and damage to the environment.

Source: Honi Brian, Cormark Communication 2002, cormark@ircradio.co.za

Sources

Bland, M (1998) *Communicating out of a Crisis*, London: Macmillan Business.

Chemical and Allied Industries' Association (1999) *Responsible Care: a Public Commitment*, Johannesburg: CAIA.

Chemical and Allied Industries' Association (1998) *Responsible Care: Management Practice Standards*, Johannesburg: CAIA

Dow (2001) *Public Report Update 2000*, Michigan: Midland.

Mitroff, I I & Pauchant, T (1990) *The Environmental and Business Disasters Book*, New York: Shapolsky Publishers.

Skinner, JC, Von Essen, LM & Mersham, GM (2001) *The Handbook of Public Relations*, sixth edition, Cape Town: OUP.

7

Crisis operations guidelines

When to activate a crisis plan

When to activate a crisis plan is perhaps the most critical question in crisis planning. Each company must determine for itself exactly what constitutes a major crisis and warrants full implementation of the crisis plan.

The key factor in determining whether to implement the plan is the presence of an actual or potential danger to employees, members of the community, the environment or the company as a whole – specifically those circumstances where company personnel do not have the usual or desired level of control.

The following circumstances are generally considered emergency situations if they present real or potential danger to employees, the public, the environment, or the company as a whole. These are offered as general guidelines.

1 fire or explosion
2 transportation accident
3 release of toxic chemicals into the environment
4 industrial accident
5 flood, cyclone, or other natural disaster
6 hostage/kidnapping situation
7 sabotage or product tampering
8 major product failure that jeopardizes lives
9 management upheaval
10 violent labour dispute
11 hostile takeover

12 criminal activity
13 terrorism
14 _____
15 _____
16 _____
17 _____
18 _____

When in doubt, it is recommended that the plan be activated and that team members be advised of the precise nature of the incident and that the plan is being implemented as a precaution.

Other circumstances that may warrant complete or partial implementation of this plan include:

1 any of the above incidents occurring at an industrial site in the vicinity, or anywhere in the local community, that could impact on site operations or employee safety
2 a notable crime that occurs on site or is linked to an employee
3 a public protest against the company
4 a potentially damaging rumour
5 impending legislation that could jeopardize business
6 _____
7 _____

Assessing priority actions

Fast and decisive thinking is the hallmark of effective crisis management. To aid in decision-making, it is important that all members of the response team understand the crisis 'decision tree'. The following is a list of priorities of what should be addressed, in approximate order, during site emergencies. These priorities should serve as decision-making guidelines for the emergency management response team.

1 matters of life and death
2 imminent harm to employees or the general public
3 short-term public health and safety concerns
4 long-term public health and safety issues
5 environmental issues
6 community property
7 disruption of local communication or utilities

8   disruption of local traffic or transportation systems
9   company reputation
10 company property
11 disruption of company business operations or customer service
12 disruption of local community business operations

Activating the network

The highest ranking manager (who could also serve as the crisis manager) must approve the activation of the crisis plan. The crisis manager (or his or her backup) has the sole authority to activate the communication network.

Communicating with 'need-to-know' audiences

It is imperative that all 'need-to-know' audiences be communicated with in a timely manner in accordance with the priority of the moment. The following matrix is designed to help you identify primary audiences, designate the personnel responsible for communication, and select the most expedient vehicles for communicating with the audiences before, during, and after the crisis:

Immediate (within minutes of the incident)

AUDIENCE	RESPONSIBLE PERSON	VEHICLE
1 On-site employees	_____	Alarm
	_____	PA system
	_____	Phone/cell/e-mail
	_____	Meetings
2 Local emergency personnel	_____	Phone/cell/e-mail
	_____	Alarm
	_____	Beeper
3 Community	_____	Alarm
	_____	Radio/TV

4 Community leaders   _____   Phone/cell/e-mail

5 _____   _____   _____

6 _____   _____   _____

7 _____   _____   _____

Less immediate audiences

AUDIENCE	RESPONSIBLE PERSON	VEHICLE
1 Local officials	_____ _____ _____	Phone/cell/e-mail Personal visits _____
2 Corporate management	_____ _____ _____ _____	Phone/cell/e-mail Facsimile E-mail _____
3 Families	_____ _____ _____ _____	Personal visits to homes Hot line Radio/TV _____
4 Union leaders	_____ _____ _____	Phone/cell/e-mail Personal visits _____
5 Customers	_____ _____ _____	Phone/cell/e-mail Facsimile E-mail
6 Media	_____ _____ _____ _____	Phone/cell/e-mail Facsimile On-site interviews _____

Ongoing communication (during crisis)

AUDIENCE	RESPONSIBLE PERSON	VEHICLE
1 Media	_____	Press briefing
	_____	Press releases
	_____	Electronic media distribution service
	_____	Phone/cell/e-mail
	_____	_____
2 Employees	_____	Website (intranet/extranet)
	_____	Bulletins/newsletters
	_____	Letter to homes
	_____	Bulletin boards
	_____	Workplace meetings
	_____	Advertisements
3 Customers	_____	Facsimile
	_____	Letters/mailgrams
	_____	Sales personnel
	_____	Advertisements
4 Community	_____	Media
	_____	Newspaper ads
	_____	Special town meetings
5 Shareholders/Investment community	_____	Media
	_____	Financial wire services
	_____	Letters
6 Other company management	_____	Facsimile
	_____	Phone/cell/e-mail
	_____	Personal visits
7 _____	_____	_____
8 _____	_____	_____

Follow-up communication

AUDIENCE	RESPONSIBLE PERSON	VEHICLE
1 Community	_____ _____ _____ _____	Media Newspaper ads Presentations to civic groups _____
2 Employees	_____ _____ _____ _____ _____	Letters Workplace meetings Internal publications Video Letters to homes _____
3 Families	_____ _____	Personal visits Community meetings _____
4 Customers	_____ _____ _____ _____	Letters Business media Trade media Trade ads _____
5 Shareholders/Investment community	_____ _____	Financial media Analyst briefings _____

We refer you now to the next chapter which presents a detailed disaster management scenario.

Sources

Skutski and Associates & SPS Communications (1990) *Crisis Communications Planning Guide*, New York: Skutski.

Solomon, CM (1994) *Emergency Plans Begin with the Basics*, Personnel Journal, April, Vol. 73(4):78

8

Disaster management plan

Umbogintwini Industrial Complex Durban South, KwaZulu Natal, South Africa

1 Introduction

The Umbogintwini Industrial Complex (UIC) is an area of industrial land lying to the south of the industrial complex of Prospecton in the Durban South basin, five kilometres from the busy Durban International Airport and adjacent to Umlazi, one of the largest former African townships in South Africa. A number of independent companies operate at the complex and Umbogintwini Operations Services (Pty) Ltd acts as landlord for the complex.

Many of the companies are chemical industries, and whilst high safety, health and environmental standards are followed, the potential exists for incidents to occur as a result of accidents, plant problems or sabotage. Where the impacts of such incidents are contained within the confines of the UIC, they are dealt with through the implementation of the Umbogintwini Operations Services' (UOS) emergency procedure (EMPRO). If the magnitude of the incident is such that the impact zone extends beyond the confines of the UIC into the adjacent communities, it could constitute a major incident that this plan seeks to address.

Figure 10
Aerial photo of
Umbogintwini
Industrial Complex

2 Affected area

The area is defined as the area external to the UIC which has been or could be impacted on as a result of an incident. The severity of the impact on any area within this impact zone will determine the level of response required from the relevant services.

3 Course of action to be taken

Owing to the speed with which an incident could impact off site, the EMPRO Controller and the Fire Services Incident Commander are responsible for carrying out an immediate assessment of the situation, for determining the potential risk to surrounding communities and for initiating a Joint Operations Centre (JOC) if they deem it necessary.

4 Off-site actions are likely to be confined to two main categories:

4.1 Sheltering

Individuals in the community may need to be sheltered in response to an incident affecting off-site areas. This entails going indoors, closing windows, not using air conditioners and monitoring the local radio station for updated information until the danger has passed.

4.2 Evacuation

The evacuation of a community as a response to an incident affecting the off-site areas will only be initiated if a delay factor exists or if sufficient early warning occurs.

5 Aim

To minimize the injuries, loss of life, damage to property and the environment arising from an incident within the UIC impacting on the surrounding communities.

Execution of plan

6 General outline

The incident/disaster management system that will be utilized to ensure a rapid and efficient response to an incident in order to minimize the impact and bring about normalization is a three-tiered system viz., operational, tactical and strategic decision making, resource deployment and support.

6.1 Operational level

The specific company where the incident originates will implement its own 'on-site' emergency response procedures. This will include the activation of the UIC EMPRO system. This, in turn, will result in the summoning of assistance from the appropriate Emergency Services.

The EMPRO structure will provide for early on-scene co-ordination of remedial actions, the establishment of forward control post(s) (FCP) and the issuing of initial instructions regarding public warning, sheltering advice and public safety information.

In conjunction with the specialists that form part of the EMPRO structure, the EMPRO Controller and the Fire Services Incident Commander will determine whether the activation escalation is necessary.

6.2 Tactical level

This level of actions will be characterized by the formulation of a JOC. The multi-disciplinary inputs will be coordinated by the Metro Fire and Rescue Services in order to determine priorities, deployment of resources and the issuing of further public safety notification. It will consider remedial/response measures related to both on-site and off-site issues.

6.3 Strategic level

On the recommendation of the JOC, the Chief of Disaster Management (CODM) or his representative may initiate a strategic level response/command in order to establish a framework of policy within which the Tactical Commander(s) will operate. This will be characterized by the activation of an Emergency Operations Centre (EOC) at which the Strategic Command Group will determine plans for the return to a state of normality.

Its primary focus will be to minimize injury and protect life, property and environment relative to the off-site impact of the incident.

7 Key functions of internal agencies

7.1 Metro Police

7.1.1 Cordon off roads and control access to and from the affected area.

7.1.2 Issue warnings and instructions to the public.

7.1.3 Assist with crowd control on an ad-hoc basis in support of SAPS.

7.1.4 Assist with search and rescue operations in support of SAPS.

7.1.5 Assist with the evacuation of people in conjunction with the SAPS and SANDF (in support of SAPS).

7.1.6 Provide a senior officer for duty at the JOC.

7.1.7 Establish and control a vehicle-holding area.

7.2 Metro Health/Environmental Department

7.2.1 Carry out full environmental survey of affected area.

7.2.2 Issue warnings and instructions to the public.

7.2.3 Assist with the evacuation of people, if necessary.

7.2.4 Provide a Senior Officer for duty at the JOC.

7.3 Health Department

7.3.1 Establish mass care centres. This will include provision of emergency housing, food, clothing, water, ablution facilitation and primary health care.

7.3.2 Establish health services aimed at individuals with special attention to the supervision of the handling of food, provision of water and the prevention of nuisances.

7.3.3 Counsel the public regarding action before, during and after states of disaster in respect of public health.

7.4 Metro Fire and Emergency Services

7.4.1 Carry out fire fighting and rescue services as dictated by legislation and internal procedures.

7.4.2 Provide a Senior Officer for duty at the JOC.

7.4.3 Provide a representative at the EOC.

7.4.4 Determine resources mustering points in conjunction with relevant emergency/essential services.

7.4.5 Make available the facility to host the JOC/EOC.

7.5 Metro Physical Environmental and Civil Services

7.5.1 Provide heavy plant equipment as required.

7.5.2 Provide transport for the movement of persons from the affected area to a safe place.

7.6 Metro Public Relations Department

7.6.1 Establish a media coordination strategy in accordance with the Disaster Management Committee requirements.

7.6.2 Liaise with the media and conduct press conferences and releases as required.

7.6.3 Assist local authorities with information control in order to prevent rumours, panic and clarify issues on hand for the printed and electronic media.

7.6.4 Have close liaison with UIC public relations staff.

8 External agencies

8.1 Airports Company

Will divert air traffic.

8.2 Ambulance and Emergency Medical Services

8.2.1 Act in accordance with medical protocol.

8.2.2 Provide senior representation at EOC.

8.2.3 Provide senior representation at the JOC.

8.2.4 Initiate and coordinate aerial evacuation of patients in conjunction with SAAF.

8.2.5 Control employment of Private Ambulance Services.

8.3 Metro Rail

8.3.1 Notify rail traffic of the incident.

8.3.2 Stop rail traffic where required.

8.3.3 Respond to commuter emergencies.

8.4 Road Traffic Inspectorate

8.4.1 Control access to and egress from disaster area on National Routes.

8.4.2 Assist in the determination of best alternative routes to and from the disaster area.

8.4.3 Assist with the dissemination of warning and instructions to the public on National Routes.

8.4.4 Provide senior representation at the JOC and EOC.

8.4.5 Assist the Metro Police in their function.

8.4.6 Assist with public warnings and public address messages (broadcasting).

8.5 South African National Defence Force

8.5.1 Give assistance with regard to the movement and protection of people and property.

8.5.2 Support the SAPS where required.

8.5.3 Provide senior representation at the JOC and EOC.

8.5.4 Activate South African Medical Services to act in support of Ambulance Emergency Medical Services (AEMS).

8.5.5 Provide 'ground-shout facilities' when required.

8.5.6 Assist SAPS with the protection of evacuated areas.

8.6 SAPS

8.6.1 Assist with public warnings and the control of public movement.

8.6.2 Cordon off affected areas and protect evacuated areas with SANDF support.

8.6.3 Provide senior representation at the JOC and EOC.

8.6.4 Carry out crowd control, search and rescue, and evacuation of affected areas.

8.7 Spoornet

8.7.1 Notify rail traffic of the incident.

8.7.2 Stop rail traffic where required.

8.7.3 Be prepared to respond to rail tanker emergencies.

8.8 Umbogintwini Industrial Complex Control Group

8.8.1 Implement UIC (EMPRO).

8.8.2 Provide technical expertise/input to the JOC and EOC.

8.8.3 Provide chairmanship for JOC in terms of National Key Point Act.

9 Coordinating instructions

9.1 Timings

As an emergency contingency plan, pre-determined timings are not possible. All actions should be carried out swiftly once ordered.

9.2 Routes

9.2.1 Fire and rescue routes will be dictated by the prevailing meteorological condition.

9.2.2 The initial call for AEMS from the UIC EMPRO Controller is to specify the approach route depending on the meteorological conditions at the time.

9.3 Central Holding Area

9.3.1 Vehicle-holding areas will be established by the relevant authoritites.

9.3.2 Metro Police will control the vehicle-holding area.

NB: The prevailing meteorological conditions will determine which vehicle-holding area will be activated.

9.4 Forward Control Post

The FCP will be established by the first emergency services arriving at the scene of the incident.

9.5 First Aid Post (FAP) and Casualty Clearing Station (CCS)

These will be determined at the FCP and communicated to the JOC.

9.6 Joint Operations Centre

The JOC will be established at one of the predetermined points, depending on the meteorological conditions.

9.7 Emergency Operations Centre

The EOC will be established at one of the predetermined points, depending on the meteorological conditions.

9.8 Helicopter landing zone

Predetermined Helicopter Landing Zones (LZ) have been declared at Umbogintwini. Demarcation of the landing zone is to be carried out by UOS Security in conjunction with AEMS.

9.9 Meteorological changes

The EMPRO Controller obtains regular meteorological updates for further dissemination.

Administration and logistics

10 Transport

Additional vehicles are sourced from transport operators if an evacuation is contemplated.

11 Evacuee assembly points

These assembly points have been identified.

12 Mass Care Centre Activation and Logistics Support

Support venues have been identified where displaced persons are to temporarily remain until the emergency situation is over. The local authority is to provide basic short-term relief and shelter.

Command and communication

13 Command

13.1 All line functions retain command and control over their own resources and personnel.

13.2 A JOC will be the tactical coordinating body and is chaired by the Joint Planning Committee chairman or his appointee.

13.3 The EOC will be the strategic coordinating body and will be chaired by the Local Council Chief of Disaster Management/Chief Executive Officer or his appointee.

13.4 The EOC is responsible for declaring a Disaster, if necessary, and for communicating this decision to the relevant provincial and national authorities.

Communication

14 Communication with public

14.1 The public at large will be notified of an incident via the UIC EMPRO Alarm.

14.2 Incident with immediate local- and off-site impact

14.2.1 EMPRO Alert; siren with undulating note; duration, five minutes.

14.2.2 All Clear; siren with continuous drone; duration, one minute.

14.2.3 News updates and information via electronic media (East Coast Radio, Radio Metro, Ukhozi FM, Lotus FM).

14.3 Communication with PR

14.3.1 Representatives of the JOC and EOC will keep the Public Relations Department updated during the emergency by means of situation reports and information bulletins.

14.4 Press Conferences

14.4.1 Communication with the media will be organized at JOC and EOC level only and details released at the time.

NB: No individuals are permitted to release any information without prior consent from the JOC and EOC.

This plan is issued under the authority of the Chief Executive Officer, South Local Council and is supported by the General Manager, Umbogintwini Operations Services on behalf of the Umbogintwini Industrial Complex.

The plan was compiled by a multi-disciplinary task group comprising representation from agencies required to provide support in case of a disaster related to the UIC.

Community liaison plan

1 Purpose

This procedure defines the processes used to inform the communities surrounding the Umbogintwini Industrial Complex on the potential for incidents arising within the Complex and impacting on their communities, and the action they should take in the event of such incidents. In addition, it describes the actions taken to keep affected communities informed during such incidents.

2 Scope

It applies to the Community Liaison Officer and the South Local Council Public Relations Officer in the aspects of communication with the community and to the EMPRO information group and the Joint Operations and Emergency Operations Centres in those aspects pertaining to information disclosure during a disaster.

3 Background and references

3.1   In order to plan an effective strategy for the on-going communication with communities surrounding the complex, certain areas that are at risk have been identified, with particular note taken of sensitive areas, e.g. old-age homes, schools, day-care centres and vulnerable communities who do not have access to facilities for their protection.

3.2   Contact has been made with a number of community organizations within these areas, e.g. women's institutes and civic associations, in order to ensure that the information is disseminated as widely as possible.

4 Definitions and abbreviations

4.1   Complex refers to the Umbogintwini Industrial Complex

4.2   UOS refers to Umbogintwini Operations Services (Pty) Ltd

Disaster refers to an incident arising within the Complex, which is of such a magnitude that the impact zone extends beyond the Complex boundaries into the surrounding communities and, in the opinion of the EMPR0 Controller and the Fire Services Incident Commander, necessitates the activation of the Disaster Management Plan.

5 Responsibilities

5.1   The Community Liaison Officer (Complex) and the Public Relations Officer (South Local Council) are responsible for the continuing dissemination of education information.

5.2   The EMPRO information group and Joint Operations/Emergency Operations Centres are responsible for authorizing the dissemination of information during a disaster.

6 Procedure

6.1 Education Information

6.1.1 A leaflet containing information as described on the next page is prepared in a format and language appropriate for the targeted community:

6.1.1a   Lists of hazardous substances used or produced within the complex and the potential hazards they pose, e.g. smell, gas and fire.

6.1.1b   Information on the EMPRO Alarm (Siren)
Tuesday testing at 10:00 sounding for 30 seconds.
EMPRO Alarm: an undulating note of 5 minutes' duration
ALL CLEAR: a continuous note of 1 minute's duration.

6.1.1c   Action to be taken in an emergency:
Protection of family and pets by sheltering indoors, shutting all windows and doors, switching off air conditioners and fans until the ALL CLEAR is heard.
Broadcasting of updated information via the local radio stations, EAST COAST RADIO (94.0 FM), RADIO LOTUS (98.7 FM) and UKHOZI FM (90 FM).
Environmental Line contact number for concerns and queries.
Toll-free line: 080 131 5171
Cellular phone: 082 440 7207
Recommending that people move outdoors and open doors and windows once the ALL CLEAR is sounded.

6.1.2   Education Information Distribution

6.1.2a   The primary educators are the Community Liaison Officer (Complex) and the Public Relations Officer (South Local Council).

6.1.2b   The most common vehicles for dissemination of information are:
Schools are visited at appropriate intervals through the year to inform (and refresh) via talks and discussions.
Interest groups/Organizations such as Women's Institutes and Civic Associations are visited, informed and requested to spread the information.
On site tours are organized on request.
Newsletters and brochures can be distributed in a Knock & Drop service.

Posters may be placed in appropriate areas. Calendars carrying pertinent CAER information are issued to the public annually.

6.2 Disaster communication

6.2.1 During an incident, the information group stationed in the EMPRO Control Room will, under the authority of the EMPRO Controller/Joint Operations/Emergency, advise the community on action to take as described in 6.1.1.c.

If the incident necessitates evacuation, the community will be informed of the location of the nearest Mass Care Centres and of any items they should take with them to these centres, e.g. medication, clothing, ID.

6.3 Disaster Communication Mechanisms

6.3.1 The information group in the EMPRO Control Room will respond to telephonic queries from the public.

6.3.2 Identified schools, churches, day-care centres, libraries, hospitals, clinics and shopping centres, etc. will be contacted and informed of the emergency. Telephone lists of these organizations will be maintained in the EMPRO control room.

6.3.3 SMS messages on cellular phones can be used to reach a wide stakeholder base.

6.3.4 Where appropriate, the JOC may authorize use of 'Ground Shout' facilities (mobile public address systems).

6.3.5 When authorized to do so, the information group will inform the three local radio stations of the situation, requesting that they broadcast information provided to them.

6.3.6 All media queries must be directed to the Plant Manager of the company involved in the incident or the UOS General Manager.
NB: No unauthorized person may release information to the media.

6.3.7 Once the situation has returned to normal and the ALL CLEAR is sounded, the information group will use the above channels to communicate this to the public.

7 Records

7.1   The Community Liaison Officer retains records of education communication for at least three years. Thereafter they are disposed of via the UOS waste-paper disposal system.

7.2   The Senior Security Officer retains EMPRO related communication records for a period of at least five years. Thereafter they are disposed of via the UOS waste-paper disposal system.

Source

This chapter was supplied by Umbogintwini Operations Services.

For more information please contact the General Manager, Private Bag X501, Umbogintwini 4120.

9

Media relations

Media relations

Any major accident or natural disaster becomes a matter of public record as soon as local emergency personnel arrive. Such events are newsworthy, and the media have a right to investigate. Virtually all companies recognize their obligation to cooperate with all legitimate news organizations regarding such events.

A review of the crisis communication policies of many companies reveals considerable agreement regarding the general policies that should guide media relations during a crisis or related activity. Consider the applicability of the following statements.

Cooperation – 'We will cooperate with the media to the fullest extent possible and respond to all media inquiries in a timely manner, **even if we have no specific comment to make or have no new information to offer**. Since responses of "no comment" or "unavailable for comment" may be misinterpreted by the media or the public, we will strive to respond at all times.'

Experience shows that in time of crisis, a non-cooperative attitude can contribute to an information void, which reporters may fill by resorting to less informed sources. This can result in misleading or inaccurate information.

Timely response – 'All media work under tight deadlines. It is critical that we respect those deadlines and provide the information that is required as soon as it is practical to do so.'

Spokesperson – 'Ideally, all interviews with the media should be handled by a single, predesignated spokesperson.'

Approvals – 'All written information should be reviewed by management, public relations, and (when appropriate) legal personnel prior to release.'

Disclosure of facts – 'Our general policy will be to disclose only absolute facts as we know them, when we know them – and to avoid any and all speculation. If we do not know or do not have access to facts or technical information, we will say, "We do not know." (Again, the phrase "no comment" should not be utilised in conversations with the media.) We will not talk with the media "off the record".'

Community concern – 'Concern for the health and welfare of employees and the community must clearly be communicated as our top priority.'

Access to management – 'We will attempt to provide the media access to the highest ranking local manager for local incidents and to the CEO of the corporation or his/her designate for crises with broader implications.'

When to contact the media

Not all crises happen instantaneously or have immediate impact upon human life or the environment. As a result, it may be difficult at times to ascertain when a potentially dangerous situation (such as a broken pipe or an environmental spill) warrants community and media attention.

There are two general 'rules of thumb' that can guide decision making during such circumstances:

1 If the situation might escalate beyond the control of site management and impact on public safety or the environment, we have an obligation to cooperate with the media and convey the facts as we know them.

2 As a general rule, even if the incident is relatively minor, it is usually better to take the initiative and disclose the information – no matter how negatively it might position the company – than to withhold the information from the media and have a reporter discover it after the fact.

What to do first

As soon as the public relations coordinator learns of the incident, he or she should alert security, managers, and other on-site personnel that reporters might arrive at the site immediately. One person on site should escort the media and refer inquiries to the designated media spokesperson and/or highest ranking member of management who is present, but is not involved with emergency personnel.

The public relations coordinator should then get to the site of the emergency (if applicable) as fast as possible and gather as much factual information as possible without interfering with emergency personnel.

Names, ages, addresses, and positions of all injured people should also be noted, as well as hospitals where the injured were taken.

If there are any fatalities, a predesignated member of the emergency response team should notify next of kin. The names of fatally injured employees or members of the public should not be given to the media prior to notification of next of kin. In any case, virtually all media have policies to withhold the release of names from broadcast or print until the family has been officially notified. Notification of kin should be prioritised to avoid causing undue concern among families not affected. There should be no delay in the flow of this important information.

The public relations coordinator should also designate someone to take photographs of the scene (for general documentation and insurance purposes).

If the scene is unsafe, reporters should be advised accordingly and escorted elsewhere for interviews.

Initial contact

All initial contacts with the media by the members of the crisis team should be reported to the public relations coordinator. The initial written response by a member of the crisis team should be cleared through the crisis manager, public relations coordinator, and legal counsel, if possible. However, it is important that an immediate response **not** be withheld if the approval process would be impractical under the circumstances.

Briefing the spokesperson for interviews

News briefings

Executives scheduled to talk to the media should be provided with a background briefing in advance of the interview. This policy should be followed even if the interview is only to be a short telephone call.

If possible the brief should include the following:

- date, time and location of the interview
- name of the reporter
- name of the publication, radio station, etc.
- previous experience with the reporter or publication – to help the executive understand the degree of caution needed in this interview and to prepare for the specific reporter's approach.
- subjects/issues/questions to be covered by reporter
- the company's position or recommended response and the data needed to discuss these subjects
- top three to five messages we wish to make in the interview (not necessarily based on the reporter's suggested topics)
- list of other executives to be interviewed during this visit, including key topics and messages you suggest the other executives cover
- issues, if any, that the executive(s) should avoid, and recommendations on how to sidestep them
- background information/statistics that would be useful in preparing for the interview
- proposed length of interview

If possible, this background material should be conveyed **in writing** to the executive so he has a chance to review it carefully. Only under exceptional circumstances should you rely on an oral briefing.

In critical situations, it is also useful to prepare a thorough set of questions and answers to define the organization's position and to be used in rehearsing the executive.

Snapshot: Understanding media

It is common knowledge that some media thrive on sensationalism and controversy. Sometimes, reporters will exaggerate facts and opinions to create sensationalism. Bad news often makes the front page or prime time television slots.

In order to win your case in the media you will need to be prepared for what Bernstein (2000) calls 'The Five Conundrums of Media Relations'.

1   A reporter has the right to challenge anything you say or write, but will bristle when you try to do the same to them.

2   A reporter can put words in a naive source's mouth via leading questions ('Would you say that ...? Do you agree that ...? Do you feel that ...?')

and then swear by the authenticity of those quotes.

3   The media will report every charge filed in a criminal or civil case despite the fact that a civil case, in particular, can make all sorts of wild, unproven claims with coverage focusing far more on the allegations than on responses by a defendant.

4   The media usually carries a bigger stick than you do through its ability to report facts selectively and characterize responses. There is a public perception that 'if I saw it in/on the news, it must be true'.

5   'Off the record' often isn't and 'no comment' can be construed as 'I've done something wrong and don't want to talk about it.'

Selecting the spokesperson

The spokesperson is a key individual in crisis communication. In the eyes of the public and the media, this person represents the total company and what it stands for. Centralizing all media contacts with a single spokesperson minimizes the possibility of conflicting statements.

Careful thought should be given to designating the primary spokesperson. Depending on the situation, instead of the company scientists or expert, it may be advisable to seek out a third-party individual to act as the chief spokesperson. In mishaps involving the local community, a vendor, or a local utility, or for incidents where more than one party is involved, the best strategy may be to allow another organization to provide the central spokesperson, particularly if that person would have greater public credibility.

A good spokesperson should:
- have credible credentials
- be able to speak for the company
- be articulate

- ideally, be the person 'in charge' of the crisis
- know top management policies
- be accessible throughout the crisis
- be able to explain technical matters in lay terms
- have the ability to phrase thoughts in concise statements (30 seconds or less)
- project a rational, controlled voice
- have a community perspective of the problem
- speak on a human-to-human basis
- understand the format, deadline and editorial requirements and policies of the media
- demonstrate concern and empathy for those affected

Guidelines for spokesperson

- Do not speculate. Always stick to the facts.
- Focus on two or three key messages to communicate and repeat them during the interview. Keep answers short and to the point. Both TV and radio reporters want 'soundbites' of 10 to 15 seconds to support the 'story'.
- Use a technical expert. There is no substitute for knowledge. If the questions are outside your area of expertise, find an appropriate technical spokesperson within the company.
- Speak in simple, common terms. Avoid jargon.
- Remain calm. Do not be intimidated into answering questions prematurely. You may tell a reporter that you need to clarify an important matter before you can answer questions.
- Do not use negative language. Do not let reporters put words into your mouth.
- Consider human safety first. When human safety or other serious concerns are involved, deal with those considerations first. You can admit concern without admitting culpability. 'We are deeply sorry for what has happened and are conducting a full investigation into the matter.'
- Do not answer questions you do not understand. Ask for clarification. Occasionally, this can be used to buy time to think.
- Ignore cameras and microphones. Face the reporter and look him or her in the eyes. Don't look away at the camera.
- Only make 'on the record' statements. There are no 'off the record' statements.

- Avoid saying 'no comment'. If you don't know an answer, say so, then lead in to your messages.

Coordinating media contacts

The public relations coordinator is responsible for coordinating all external communication. Although the media spokesperson may be the only individual actually communicating with the public, individuals assigned to different 'target' groups (e.g. community relations, government affairs) should keep records of any contacts made or requests received.

Any inquiries from the national media should be brought to the attention of corporate public relations.

Snapshot: Getting legal counsel and public relations to work together: trial by media in the court of public opinion

During a crisis, a conflict sometimes arises between the recommendations of the company's legal counsel, and those of the public relations counsel. While it may be legally prudent not to say anything, this kind of reaction can land the company in public relations 'hot water', which is potentially as damaging as any financial or legal consequence. Fortunately, legal advisers are increasingly aware of this and are cooperating with public relations counsel.

As stated before, some media use sensationalism and controversy. Once you are involved in a legal case or commission of enquiry, almost every audience important to your client's business and your legal case is going to be seeing the media's version of the alleged facts. This is 'trial by media in the court of public opinion'. You need to cooperate with them to the extent it doesn't compromise your legal strategy.

If you deliver your key messages, are media trained, and view the media as a gateway to important audiences rather than as the enemy, you can optimize the results. Sometimes that just means being quoted accurately. Sometimes that means a story which looks very good for your company.

The worst 'trial by media' experiences occur when an organisation is facing emotion-eliciting charges, such as causing death or breaching the public trust. In these situations, the media tend to editorialize in the guise of reporting, pandering to the emotionalism of the public.

Clients with a weak case or limited financial resources have often engaged in deliberate 'trial by media' tactics to force a settlement, with mixed results. It's always a risky tactic because no one can reliably manage the media; but some succeed in winning through embarrassment. An organization has to think about its long-term strategy. For example, if the organization thinks it might want to attack the prosecution for improper media disclosure at some later date, it's important to take a lower profile at first.

The reality of today's sensationalist media and public environment is that regardless of the legal merits of any crisis situation, perceptions generated from the onset of the case through to its resolution can dramatically impact on the reputation and economic welfare of the organisation. Perceptions, which can be as helpful or damaging as the provable facts, can also affect the attitude of prosecutors or regulators in the legal process. The role of public relations therefore is to help stabilize that environment by developing a public relations strategy which results in prompt, honest, informative and concerned communication with all important audiences – internal and external. Strategies that defer to legal considerations to the exclusion of all else are dangerous because the organization remains 'bound and gagged' and vulnerable to those who continue to communicate negative and damaging messages about it.

In a criminal case, public relations is particularly important during the investigatory phase because there is a greater opportunity to influence how the client is viewed by the media. Ideally journalists receive a positive first impression which will carry through the investigation. Covering up or stonewalling leads to a negative impression that will be difficult to erase.

A crisis management expert can brief the legal counsel on what counsel plans to show to the judge and prosecutor, and on how this will be presented in the media. Having a professional spokesperson other than legal counsel can prevent sometimes overwhelmed attorneys from reacting inappropriately to eager reporters.

Are the following statements about a civil or criminal case true?

1 Witnesses never lie regarding their advance bias about a case.

2 Prosecutors, magistrates and attorneys never talk about a case outside of court deliberations, or read and watch TV about a case when sequestered, when directed not to do so by a judge.

If your answer is 'no' you begin to appreciate the potential value of crisis/issues management. While it is unethical to attempt to influence witnesses and legal counsel, there is nothing unethical or unprofessional about presenting an accurate picture to the media and other audiences.

Any honest reporter will admit that he or she brings a natural bias and an editorial perspective to a story. Journalists will do their best, in that context, to report in a 'balanced manner', with the exception of columnists, who are often free to say pretty much what they please. These media representatives are a gateway through which both plaintiff/prosecutor and defendant can communicate to a variety of influential publics. It is the responsibility of counsel, with expert assistance as necessary, to direct media relations that can shift the balance of coverage.

Legal experts realise that the same analysis done by crisis management professionals to anticipate multi-audience response to various public relations tactics also helps them anticipate legal response.

The crisis consultant can sit in on practice sessions for depositions, resulting in a change in the way a client expresses his or her position. Tactics used in this public education process can include:

- Educating employees about what to say and not to say about the situation when they're away from work.

- Advertorials – buying print space or broadcast time in which one puts news-like stories about your client organization designed to balance any misinformation already in the public eye. This tactic is usually only employed if the media has consistently misreported the facts.

The battle for legal decisions does not begin in the courtroom. It begins with advance communication immediately a legal battle is likely. It can work together with legal tactics to preclude a case ever going to trial (assuming that's a desired outcome for either side of the issue) or to affect public perception sufficiently to enhance either side's chance of a favourable outcome in court.

Issuing statements

The initial statement that follows the accident or incident can be the most critical, in terms of company credibility. After the situation has been assessed, a statement should be prepared and approved by the appropriate management. This statement will serve as official notification of the accident or incident and should meet national, provincial, local and regulatory agency requirements for emergency announcements. Included in the statement should be the following:

- nature of the accident or incident
- whether the public or the environment is in any danger
- the 'who, what, when and where' of the incident (be as specific as possible)
- when local authorities were notified
- what products, processes, or materials were involved
- response to the emergency
- steps taken to contain or remedy the situation (include emergency personnel contacted)
- heroic actions by employees or emergency personnel (identify)
- extent of injuries and/or deaths
- persons to contact for further information

Many companies caution against releasing the following types of information:

- speculation of any type (particularly on the cause of the emergency)
- rand estimates of damage
- information about insurance coverage
- premature assessments of the performance or reaction time of company or local emergency response personnel
- implications of negligence
- overly descriptive words in explanations, such as 'devastating' or 'careless'

Means of distribution

Every effort should be made to ensure full and complete disclosure of all written statements to all pertinent media. The following are the recommended ways of distributing media information on an immediate basis:

- phone call to local news wire service

- phone calls to local newspapers, TV, and radio stations (city desks or news desks)
- facsimile and e-mail transmittal of written statements and press releases
- paid public relations distribution service

If access to the emergency scene is limited for safety reasons, disclose the information to a few key media people, who, in turn, share the information, photographs, or video footage with other media members. Try to accommodate at least one reporter from each type of media – TV, radio, wire service and newspapers.

Pre-prepared materials

Fact sheets

A fact sheet should be developed for each site. Fact sheets should include:
- number of employees
- size of operation
- products and/or services of facility
- past safety record
- markets served
- names of key management personnel
- chemicals and/or materials on site
- contributions of facility to community

Press kits

Prepared press kits for distribution to the media could save valuable time and maintain consistency. Included in the press kits should be:
- general company background
- map of facility
- photos of facility
- photos and biographical sketches of key managers
- background data on any potentially harmful chemical or material (material safety data sheets)
- company contacts (local and headquarters)
- any pamphlets or general information about the company and its product or operations
- annual report

News releases

An emergency situation might require a news release to be distributed to the media. The purpose of the news release is to convey written information on the incident and avoid misinterpretation.

The following circumstances may require a news release:

- an accident, fire or explosion that results in serious injury, death, or considerable property damage
- a health or environmental incident that may affect employees, the surrounding community, or the environment
- a serious traffic or air accident involving company vehicles, products and/or personnel
- sabotage, abduction or extortion, bomb threats, acts of terrorism involving company personnel, products or property
- news of an incident that is likely to be known by employees or circulated in the community and create misleading impressions
- news of an event that is unusual enough to cause concern to employees, nearby residents, or community officials
- consistently misleading news reports

Writing an initial statement for release

Often, reporters will call before all the facts have been gathered. In such an instance, a simple statement acknowledging the situation is useful. The short statement avoids 'no comment' and acknowledges that the company recognizes the need to cooperate with media.

Examples are:

- 'Our company is responding to the situation (name the emergency). We have trained and experienced people on site working on the situation.'
- 'Our first priority is the safety of our employees and the public. We are gathering information, and as soon as details become available, we will inform the media.'

Writing a news release

By following a few basic principles when writing news releases, your company stands a better chance of having reporters use the information with only minor changes. Remember to consult the legal department as needed.

1 Give the most important information in your lead paragraph.

Your story competes with other news and information, so the most important point should be stated clearly in the first paragraph.

2   Answer four of the five 'Ws' – Who, What, Where and When. Explain WHAT the emergency is. Identify WHO is involved in the emergency as well as the material and equipment involved. Tell WHERE and WHEN the emergency occurred. Explain WHAT action is taken to respond to the emergency. Do not explain WHY the event occurred unless complete information is available.

3   Attribute information to a qualified source. A news release is useful only if it conveys credible information from a credible source.

4   Write remaining information in descending order of importance. If the media cuts off the bottom of your story, they will cut information that is least important to the public.

5   Explain technical points in simple language. A direct quote can add the human element to otherwise technical information and help explain a situation or event in layman's terms. Tell the real story. Avoid using language that is overly bureaucratic.

6   Be concise. A good news release is judged by the quality of information it communicates, not by its length. Stop writing when you've said all you need to.

Press conferences

If the crisis is a major one, it will be necessary to brief the media on the progress of the crisis throughout the ordeal. When a press conference is planned, careful consideration should be given to:

Location

The site of the incident may not be the best location for convening a conference. A nearby hotel, government building, etc. may be more appropriate. For a crisis of national proportions, it may be advisable to meet with the press in major centres such as Johannesburg, Durban or Cape Town, as appropriate.

Speakers

In addition to the media spokesperson, other members of the crisis communication team, as well as the plant manager, technical adviser(s), and corporate representative, should attend. Typically, at a press conference, it is rare that one person would be equipped to answer all questions. Advance planning can help determine who will answer what questions. Keep in mind, however, that whoever attends the conference should be fully prepared to deal with spontaneous media questions.

Press list

The public relations coordinator and the individual assigned to press relations should develop a media list including all local news media, as well as regional and national media if the situation warrants.

Media releases and contacts logbook

It is also highly advisable that a logbook of all statements, news releases and media contacts be kept, detailing exact time and names of media contacted.

Media monitoring

All news reports should be constantly monitored for accuracy. Corrections and clarifications should be provided to the media immediately.

On-going contacts

After the initial announcements, it is recommended that the designated spokesperson be accessible on a 24-hour basis throughout the crisis. Regular updates in the form of daily, semi-daily, or even hourly statements may also be necessary.

Evening and weekend contacts

It is critical that the media know how to get in touch with a public relations representative during weekend and evening hours. Therefore, all individuals who may receive a media call to the general site phone number must be provided with the home phone numbers of the local and corporate public relations coordinators. The standing instructions would be to contact a public relations representative immediately upon receipt of an off-hour media call.

The public relations coordinator may also want to assign 'second- and third-shift' media contact duties to other personnel during times of crisis (it must be remembered that all major media maintain 24-hour operations).

'Total story' advertorial

Some observers of crisis management argue that – given the way the media operate with deadlines and limited space and airtime – a company's story may be told in bits and fragments over a prolonged period of time, making it difficult for the public to grasp the 'total story'. It may, therefore, be useful to consider developing an information advertisement in a local or national newspaper to ensure that the public is provided with the complete story in one place, from a single source.

Media contact list

It is very important to have an updated directory of media contacts available in the event of an accident or crisis.

MEDIA	CONTACT	PHONE	E-MAIL

It is also important that a very careful record is kept of every call made to the company during the crisis period by media personnel as follows:

MEDIA MONITORING

Who called	Publication	What they asked	Deadline	Our response	When we responded

After the crisis is over it is vital to:
- continue maintaining relationships with stakeholders
- be proactive in your communication
- try to rebuild your reputation.

A balanced, well-timed public relations response in a crisis minimizes the chances of misinformation, misinterpretation of the facts and damage to corporate reputations. A sound, proactive media policy is an essential part of this strategy.

Sources

Bernstein, J. (2000) 'An ounce of prevention: The Five Conundrums of Media Relations' in *Crisis Manager: The Internet Newsletter about Crisis Management* [Online] Available:
www. bernsteincom.com/nl/crisismgr 000515. html

Bland, M, Theaker, A & Wragg, D (1996) *Effective Media Relations: How to Get Results*. London: Kogan Page.

Bland, M (1998) *Communicating Out of a Crisis*, London: Macmillan Business.

Cutlip, SM, Center, AH & Broom, GM (2000) *Effective Public Relations*, New Jersey: Prentice Hall.

Fearn-Banks, K (1996) *Crisis: A Casebook Approach*, New Jersey: Lawrence Erlbaum.

Regester, M & Larkin, J (2002) *Risk Issues and Crisis Management – A Casebook of Best Practice*, London: Kogan Page.

Skinner, JC, Von Essen LM & Mersham, GM (2001) *Handbook of Public Relations*, sixth edition, Cape Town: OUP.

10

Follow-up communication and performance checklist

Follow-up reports to 'need-to-know' audiences

Communication should continue with a number of audiences even after the incident has been resolved. For example, thank local officials and emergency personnel for their cooperation. Notify employees of the impact of the incident on business operations. The community at large will retain an interest for some time in the incident's impact upon the company, the local economy, the environment or community health.

Often, many unanswered questions or, worse yet, false rumours may persist after an incident. In addition, there may be unresolved questions of liability.

Finally, the confidence of customers, suppliers and distributors must be restored. Make every effort to assure customers that, in spite of the incident, the company is committed to addressing their needs. Give as much specific information as possible about reopening dates, shipping dates, etc. (These should be reasonable – not overstated – promises.)

It is therefore recommended that a post-crisis public relations plan be developed and implemented.

Media follow-up

If possible, an interim statement should be provided within a few hours after the event and certainly a comprehensive statement within 24 hours.

Throughout and beyond the crisis the story needs to be monitored closely and the PR staff given access to confidential information in order to respond timeously.

What needs to be stressed, however, is that it is not only a question of right or wrong, but that the credibility of both individuals and the organization are at stake.

The media, in particular, will be interested in obtaining follow-up information such as:

1   statements on subsequent aid to the community, remedial or clean-up actions, etc
2   statements on liability
3   statements on business impact
4   information on past safety records and related incidents
5   information on other incidents at comparable facilities within the company or the industry
6   local reactions to how the company handled the crisis (including investigations into whether industry safety or emergency procedures were properly followed)
7   follow-up on injured persons
8   results of investigations into causes

Spokespeople should anticipate any questions that might arise in the aftermath of an unexpected incident and prepare appropriate responses.

Employee communication

The site manager should be open and candid about the incident with all employees, explaining the event, the company response, and the precautions taken to prevent a repetition.

Assign a special task force to monitor employee reaction and any unexpected employee relations problems resulting from the incident. Experience has shown that, in times of crisis, employee issues and concerns might be unintentionally overlooked because of other priorities.

Letters to employee homes, workplace meetings, and articles in company publications are all appropriate vehicles for employee communication, as are one-on-one meetings, bulletin board postings, and presentations by external officials involved in the crisis.

Customer communication

The crisis team member responsible for customer relations should assess the business impact of the incident and determine the best means for communicating with customers, particularly if the incident proves to be disruptive to normal business operations.

E-mails, facsimiles, and, of course, direct phone calls are appropriate means for immediate communication. Follow-up letters or e-mails are the best means of dispelling rumours and conveying other important information (such as special arrangements for handling pending orders) to customers after the incident.

A major consumer problem might require extensive television, radio and newspaper advertising, a media tour and other mass-media techniques. For example, in any product recall, the guiding principle must be the well-being and safety of the consumer. That should rule all decisions. It is also important that accurate information is available about the problem – that, in itself, can be reassuring because it implies that the company knows what it is doing. The terms in which any product announcement is couched are also important.

The announcement should:

- describe the problem
- tell consumers how to identify the product affected
- explain how the situation arose
- accept responsibility
- set out the action the company is taking to rectify the matter
- advise consumers if they require more details or reassurance

In any product recall, time is of the essence. A well-prepared company will rise to the occasion and deal with the problem while continuing to run its business. A complacent company will crumble under the spotlight of media and consumer attention and be in danger of going out of business.

Industry communication

Depending on the extent of the crisis, it may also be necessary to mount a targeted marketing information campaign to the industry at large. This may include a direct mail campaign, advertisements in trade and business publications, feature stories in similar publications, presentations to industry groups, etc.

Crisis communication performance checklist

While there are no hard and fast rules that apply to all crisis situations, 'post-crisis analysis' by public relations experts – in the aftermath of recent industrial accidents such as oil spills and chemical plant explosions – suggests a number of guidelines for successful crisis management. Use this checklist below to evaluate your crisis communication response – or to ensure that your company's crisis communication plan has taken key factors into consideration:

- **Immediacy:** Response to the problem is immediate. People most affected by the crisis are communicated with on a priority, need-to-know basis – in a timely, orderly fashion.
- **Concern:** All communication demonstrates that the health and welfare of those affected by the crisis are the top priorities of company management.
- **Control:** Information about the status of the crisis situation is issued from a central company control point – ideally, through a single media spokesperson.
- **Factual:** All comments made for public record are based on facts, not speculation.
- **Management:** Top management assumes an involved and visible role.
- **Solution:** The company is positioned early on in the crisis as part of the solution. Avoid an adversarial or defensive posture.
- **Apology:** For situations where it is absolutely clear that the crisis or disaster was caused by the company, a public apology is issued from a high-ranking official.
- **Access:** The media and public officials have direct access to company decision makers.
- **One-to-one:** Communication from the corporation is handled on a human-to-human level, not as if it is coming from an indifferent corporate entity.
- **Perception:** Public perceptions of the crisis are addressed seriously, regardless of whether these perceptions are based on fact.
- **Credibility:** In the selection of the spokesperson, consideration is given to the credibility and trustworthiness of the source. Third-party spokespersons are used when appropriate.

- **Advocacy:** The company's right to communicate is fully exercised and not relinquished to other parties involved in the controversy. If the media and public don't hear the story from the company, they'll get their information elsewhere.
- **Diplomacy:** In controversial situations, efforts are made to sit down with adversarial groups to initiate dialogue and negotiate solutions.

Snapshot: How to run a successful community safety and information campaign

A chemical company operating in a built-up area on the East Rand, not far from Johannesburg, recently ran a community information campaign, under the auspices of its Community Awareness and Emergency Response (CAER) committee. The committee comprises community and civic leaders and other interested parties from nearby communities, safety personnel from adjacent businesses and industries, representatives of the local Emergency Services and members of the Environment, Health and Safety (EH&S) department at the plant.

The campaign was conceived as a public awareness exercise intended to promote safety in communities living and working around the manufacturing site.

Purpose

The primary objective was to inform community stakeholders about measures to take to protect themselves, their families and co-workers, in the event that a chlorine gas leak at the plant could not be contained and affected the surrounding community. The secondary objective was to generate goodwill by telling audiences about the company's operations and products generally, and more specifically the benefits to modern everyday living that make chlorine an indispensable commodity, as well as the comprehensive EH&S policies and procedures that are implemented to minimize the potential for chlorine gas leaks and their harmful effects. A third objective was to inform community stakeholders about the activities of the CAER committee with a view to inviting additional stakeholder representation.

Target audiences

For purposes of manageability, phase 1 of the campaign was run for people living and working within a 2,5 km radius around the site, with the planned intention of extending it outwards in concentric circles during further phases. It was estimated that complete coverage of the local community area that was likely to be affected would take some six months.

Stakeholder or special interest groups were identified as:
- residential communities
- businesses and industries
- educational institutions
- crèches, nursery schools and after-school care centres
- clinics and hospitals, medical waiting rooms
- homes for the aged
- shopping centres

in addition to:
- company employees
- emergency, safety and security service providers
- local media
- local government representatives

The site is surrounded by residential areas that cover

the spectrum from category A income levels to informal settlements, with small and large industrial and business concerns in close proximity.

The challenge was to find the appropriate combination of communication elements and channels for this very diverse audience to ensure that the messages were received, understood and retained by as many people as possible. In addition, budget cuts by the parent company during the planning phase meant that the programme had to be low-budget.

A consultative meeting with representatives from the community, local organisations and the local Emergency Services had a two-fold purpose: to enlist the cooperation of these representatives in the implementation of the campaign, and to get their input on the most appropriate means of communicating to their constituencies.

On the advice of representatives at the meeting, a number of early suggestions was discarded. The South African Police Services (SAPS) representative, for example, appealed to the organizers, as a crime prevention measure, not to distribute information material to motorists at traffic lights and stop streets.

The key channels and tools used for disseminating and reinforcing the central messages were:

- internal company media for employees
- a fun event roadshow
- brochures in the four main local languages
- advertorial and editorial in the local community newspapers
- posters
- buckets with reminder labels distributed to learners at local schools
- small group meetings to inform service providers, such as clinic personnel
- training sessions with local school teachers
- one-on-one information sessions with the stakeholder groups identified above

It was realized in the early planning stages that a pub-

lic meeting of residents would not be effective. Instead, it was decided to launch the campaign by a roadshow presented to parents and pupils at the two largest local schools. The information content met the objectives listed above, and was presented by way of demonstrations and role-play representations of emergency response procedures and measures that individuals should take to safeguard themselves, by presenters from the company, the Emergency Services and the SAPS. Audience participation helped to reinforce the key messages, and the presence of the company's fire engines and fire officers added to audience enjoyment. A festive atmosphere was created with balloons and banners featuring a cartoon character specifically designed to brand the campaign. Translators were also available.

The procedure that individuals should follow in the event of a gas leak was repeated in brochures produced in English, Afrikaans, Zulu and Sotho. These were printed on bright yellow, the campaign colour, to ensure that they would be easy to find in the future. The pamphlets were also distributed house-to-house throughout the residential areas within a 2,5 km radius.

The safety procedure for individuals was announced on bright yellow posters, which were distributed to schools, shopping centres, clinics, etc. for permanent display on noticeboards.

Given the importance of the bucket as a household tool in most homes, buckets with bright yellow labels containing the personal safety procedure were distributed to learners at schools in the area. The schools have undertaken to incorporate the topic of chlorine, its benefits and hazards, and concomitant safety precautions into their lesson content.

Ongoing information dissemination has taken the form of small group meetings with the staff of municipal clinics; training sessions for the local school teachers to use the material in the classroom; one-on-one information sessions with safety officers of adjacent businesses who have been asked to disseminate the information to their own employees; with ward representatives; crèche and nursery school owners; old age home managers; and the managements of shopping and medical centres.

Results

While no survey has been conducted to measure results, indications are that the messages have been understood and retained, particularly by learners at the local schools. This is significant as they can disseminate the information within their homes. Community leaders also report that there appears to be greater awareness of the company and its activities within their areas, as well as appreciation for the fact that the company took the trouble to 'come and speak to them'. There have also been a number of requests for the company to address them again. This is particularly encouraging, since it reflects the opening of channels of communication.

To ensure that the progress is not lost, a reinforcement programme has been recommended for communities that have already been exposed to the campaign information. This will run in parallel with the new launches that will be made to communities located within the next 'concentric circle'.

Information supplied by Honi Brian, Cormark Communication, cormark@ircradio.co.za

11

Checklists

This section contains some of the checklists we believe may assist in key aspects of crisis management. They can be used on the basis of individual corporate plans.

1 Disasters and catastrophes

Disasters and catastrophes	Assigned to	Date/time assigned	Date/time completed
1 Implement contingency plan			
2 Assign PR and 'borrowed' personnel for each specific task.			

3 Establish HQ (headquarters), field station(s) and press rooms.
3.1 Inform and clear with community leaders.
3.2 Inform and set up liaison with:
  • police
  • fire department
  • Red Cross and civilian defence
  • hospitals
  • government agencies			

Disasters and catastrophes	Assigned to	Date/time assigned	Date/time completed
4 Aggressively gather all information on nature of catastrophe, victims, survivors, damage, property damage, relationships, etc. and notify pertinent people. 4.1 Assess impacts on ancillary people. 4.2 Cover and obtain essential factors for news points			
5 Prepare news release/company statement. 5.1 Include all known facts. 5.2 Be certain that all information is accurate. 5.3 Inform next of kin (withhold names of victims until this is done).			
6 Clear statement/release with: • senior management • legal department • HR department • union(s) • police, fire, regulatory agencies, community leaders			
7 Obtain pertinent quote from CEO (designee) and insert in release.			
8 Issue release immediately to: • general-interest local and national electronic and print media • pertinent trade and industry publications • weeklies and monthlies • employees, by bulletin boards, phone and e-mail networks • community leaders • insurance company • pertinent governmental agencies.			

Disasters and catastrophes	Assigned to	Date/time assigned	Date/time completed
9 Set up field press room(s) and the 'Crisis Communication Centre'. (See diagram, page 63) • Select site that is isolated from restricted areas and away from scene of immediate accident. • Notify news media of location of press room. • Arrange telephone lines and switch-ins from PR department to press room.			
10 Assign sole spokesperson(s) who will be on duty day or night: • at site disaster • at HQ • at different liaison points • one 'roamer' to make rounds and report continuously via walkie-talkie/cell phone to HQ PR office			
11 Arrange for switchboard to route all pertinent calls to HQ PR office.			
12 Direct company employees to refrain from making any statement to media people. • Seek cooperation and consensus with union leaders on messages and statements.			
13 Inform the following that only official spokesperson will issue reliable information: • customers • dealers • vendors • shareholders			
14 Arrange for news conference as soon as CEO has prepared statement. Arrange for CEO interviews with individual press, as requested.			

Disasters and catastrophes	Assigned to	Date/time assigned	Date/time completed
15 Arrange for company photographer to obtain complete coverage for legal, file and news camera people on demand. Arrange for other pertinent interviews, photographs, and TV/radio coverage. Arrange for biographies and photos of pertinent employees to be available in press rooms. Obtain, clear and issue updated, factual news on a continuing basis to all press present and by telephone to others.			
16.1 Keep news media away from victims, survivors and relatives until given approval by appropriate persons. 16.2 Arrange for separate transportation and quarters, if necessary, for victims, survivors and relatives. 16.3 Ask hospitals to assign one individual who will notify you of significant information about injured.			
17 Remember that the sooner you get all the news out, the sooner the catastrophe ceases to be a major news event. Get all the facts. Get all the facts out.			

A special issues management checklist is provided in the second edition of Michael Regester and Judy Larkin's book *Risk issues and crisis management: a casebook of best practice* (2002), published by Kogan Page (London) in conjunction with The Institute of Public Relations (UK).

2 Preparing for emergencies

Preparing for emergencies	Assigned to	Date/time assigned	Date/time completed
1 Develop a 'worst case' list by asking the following to imagine the worst thing that could happen to the organisation: • CEO • department heads • union leaders • community leaders • police and fire departments in each plant community			
2 Review news files and clippings of past emergency situations in your industry. Consult other organisations in your community.			
3 Compile and consolidate lists for a master list. Categorise items into groups for comprehensive contingency plans.			
4 Begin with the constants which will apply in all cases. These must include: 4.1 Assignments for PR staff. 4.2 Listing of all organisation personnel to be notified, in order of priority. 4.3 Create action plan for PR and alternate plans: • establish HQ staff and tasks • establish field staff and tasks • establish liaison staff for police, hospitals, etc. 4.4 Arrange and get approval for CEO's actions and statements in emergency. 4.5 Assign organization spokesperson and alternates of 24-hour, seven-day duty. 4.6 Print basic plan, to be displayed prominently in all departments.			

Preparing for emergencies	Assigned to	Date/time assigned	Date/time completed
4.7 Explain the order in which people must call others on the notification list. 4.8 Station lists must be simple, direct, easy-stepped, and quickly implementable. 4.9 The PR staff list will be the most comprehensive. 4.10 Prepare alternate plans to fit a variety of crises. 4.11 Remember peripheral impact areas, e.g. the advertising department, which must have a substitute ad campaign ready for use in an emergency.			
5 Prepare and get approval for 'time-gainer' responses for executives. These are to be used only until a real and factual response is readied and released.			

3 Ambient press hostility

Ambient press hostility	Assigned to	Date/time assigned	Date/time completed
1 Compile, over a sixty-day period, in headquarters, plant, facility, or community: • all news stories about organization • references to your organization in other stories • record of telephone inquiries by news media • record of personal meetings with media • record of TV/radio interviews, panel shows, etc. by company personnel • record of taped or live quotations about the company			
2 Compile, during the same time period, an assessment of news media attitudes toward your organization.			

Ambient press hostility	Assigned to	Date/time assigned	Date/time completed
3 Compile, in the same period and areas: • a list of all media requests • a list of completed responses and satisfactory answers • a list of uncompleted responses to inquiries or requests – with reasons for failure to satisfy			
4 Determine a 'bottom line' attitude as a result of the above lists. 4.1 Evaluate attitudes on the part of the media. 4.2 Make an honest and factual determination whether the hostility is: • real or imagined • normal adversarial relationship • a result of personality • common to all contacts 4.3 Analyze whether the hostility is caused by the company and, if so, how.			
5 Determine corrective measures to be attempted: 5.1 Choose and implement a response to media: • gradual, and soft-sell • bold, direct confrontation to clear the air and reach a working relationship agreeable to both media and your company • disregard and live with it • defiance 5.2 If the decision is to work out some positive relationship: • Visit each news media and meet with executive and pertinent individuals; explain your desire to establish a working procedure agreeable to company and media. • Invite individual media to a business lunch to discuss relevant issues. • Have meetings with each media representative as quickly as possible. • Invite complaints at these meetings. • State company complaints. • Set up mutually agreeable ground rules.			

4 Media conferences

Media conferences	Assigned to	Date/time assigned	Date/time completed
1 Determine: • actual need • specific goal • product, service or subject • time schedule • target audience • location for conference			
2 Decide on conference site			
3 With CEO and pertinent senior officers, determine: • who will be the company spokesperson(s) • content of press kit • graphics or visual aids • list of invitees			
4 Write draft of principal statement and get clearance from: • CEO • legal department • marketing department • finance			
5 Prepare potential questions and answers to be expected from attending news media.			
6 For non-emergency press conference: • Arrange a venue. • Send out invitations. • Give advance release, two weeks prior, notifying news conference's date and time. • Phone those not heard from three days before the conference.			
7 Determine a.m. or p.m. practicality to meet media needs, depending on deadlines.			

12

Stress management

Disasters and crises generate very high levels of stress – not only in victims, but also in helpers, other stakeholders and even bystanders. In this chapter, the phenomenon of stress reaction will be addressed; first, from a crisis management perspective. Then a broader approach to understanding stress and its management in normal circumstances, crises and disasters will be described. Finally, certain coping strategies which can be applied at organizational, managerial and individual levels will be proposed.

The crisis management perspective

Effective crisis management is not primarily a set of tools and mechanisms to be implemented in organizations, but rather a general approach and set of actions by managers who are not too emotionally bounded. Essentially, crisis managers need to be able to put themselves in the position of all stakeholders and be able to see what impact the crisis may have on their feelings and needs.

Therefore a major skill when dealing with crisis is the ability to anticipate and deal with the stress experienced by others – and also to deal with the stress that you are under yourself at the same time.

When selecting a crisis management team look for evidence of these particular skills and attributes:
- clear-headed and rational managerial decision-making
- consideration of stakeholder interests in the long term, not just the short term

- ability to involve others in the decision-making process
- clear communication skills
- not being defensive
- the ability to admit one does not know all of the answers
- keeping on top of the news – reacting in a speedy fashion, but not as a panic reaction
- capacity for using available resources
- the ability to work under extreme time pressure
- the need to liaise with others and work as part of a team
- the need for flexible and creative thinking, and
- the ability to manage stress in oneself and in others (this is important in order to be able to display the skills outlined above)

Approaches to change management usually begin with the assumption that change is implemented and takes place within a pre-ordained time-frame (for example, the implementation of new technologies, downsizing). In crisis management, on the other hand, many crises and disasters are unexpected and may lack a clear beginning and end. However, change management models are helpful in that they predict what responses will occur and suggest the type of actions necessary to deal with them, as illustrated below.

Situation	Problem	Solution
Personal impact	Raises a whole series of questions: Am I to blame? What does this mean to me? Do I still have a job?	Allow all parties involved to air their views.
Resistance to change	People may decide to resist the changes imposed by the crisis.	Find out what underlies resistance. Identify people's needs during the management of the crisis process.
Readiness	People accept what needs to be done – but could experience uncontrolled enthusiasm.	Encourage greater readiness throughout the organization before a crisis occurs
Power/political dynamics	People will want to side with the 'winners' and have a say in any possible outcomes.	Identify who the political players are, know the key stakeholders and how to manage their reactions
Need for control	Because of uncertainty an individual's need for control may emerge.	Stay in control of the on-going business and control how the crisis is managed to minimize disruption
Cultural impact	A change or crisis may impact on people's feelings, values, beliefs, norms and rituals.	Conduct a cultural impact assessment which will reveal the extent of these factors as well as how committed people are to the organization.

Individual reaction to change

People attempt to handle dramatic negative changes, such as bereavement, divorce, major illnesses or job loss, by responding in the following way:

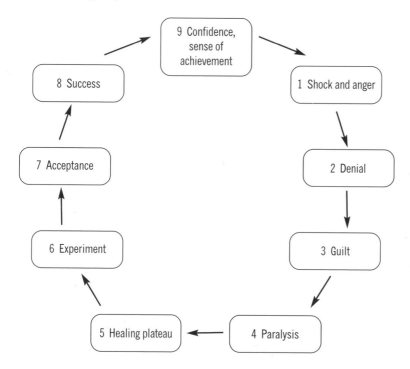

Figure 11
Individual's reactions to change

These are some of the common symptoms experienced and questions asked:

1 **Shock and anger:** Is this real?
   Why is this happening to me?
   How could they do this to me?
2 **Denial:** This isn't true!
   This isn't really happening to me.
3 **Guilt:** It's all my fault.
   If only I hadn't.
4 **Paralysis:** I simply can't handle this.
   I'm going to shut my eyes and wait for it to go away.
5 **Healing plateau:** Well, maybe it'll be OK.
   We can only go forward.
   I need some time to think about it.
6 **Experiment:** I suppose we could try this or do that.

7 **Acceptance:** I think we can make it.
   I can cope – this isn't so bad.
   Actually things may be better because of this.
8 **Success:** I managed that really well, people still trust me to do
   a good job.
9 **Confidence and a sense of achievement:** I believe I am better
   equipped now to deal with difficult situations in future.
   I am really proud for getting through this.

It is important that individuals are allowed to move through each
of these stages, but they should not be allowed to stay in each stage
for too long. They must be encouraged to move on through reassurance and the use of appropriate questions and answers.

Thus some of the main tactics that one can use to handle a
crisis successfully are:

- **enhanced awareness**
  This will help one deal with both the situation and one's own
  and others' reactions.

- **self-preservation**
  It is important that we develop a lifestyle that makes us robust
  to stress all of the time, but especially in a time of crisis.

- **physical activity**
  This helps dissipate anger, restore balance and also gives one
  some time alone to focus and concentrate.

- **relaxation**
  Some examples are yoga, sport, exercise, meditation, massage
  or mental relaxation.

- **seeking social support from colleagues, friends and family**
  Often in a crisis individuals withdraw. One needs to be able to
  identify who one can turn to, both inside and outside of work.

Selection of a crisis team

In selecting and helping individuals to function effectively in the
crisis team you should be looking for individuals who:

- have an internal 'locus of control' – that is, they believe that
  they, rather than chance or other people, are responsible for
  shaping the outcome of events in their lives.

- are not emotionally bounded, who can discuss and question aspects of the crisis openly and without defensiveness
- have positive self-regard
- are hardy – committed, in control and confident
- are not neurotic, and
- are good team members

These qualities can be identified via self-assessment and observation of teams working in training sessions. It is also important to have contact names for those in the caring and support industries, for example, counselling services, and to have information packs available.

In the emergency services, the use of standard response procedures has been found to assist in providing a feeling of control in a challenging environment. This should also apply with regard to organizations. Thus, if an organization has in place a (workable) crisis manual and a crisis team and has considered the best way to deal with a variety of crises, then individuals involved should have an enhanced feeling of being able to cope, of being able to implement some initial actions which will relieve stress.

Snapshot: The value of debriefing

There are invariably lessons to be learned from any incident. This applies to communication procedures as much as it does to the causes of the incident and the way in which it was managed. Once a crisis has been brought under control and normal operations resume as things move into the recovery phase, it is vital to review the procedures that were employed and the resources used. Debriefing assists managers to learn from the causes and resolution of the incident, providing information that can help prevent an incident in the future or minimise hitches in dealing with it if another incident should arise; and to improve on the measures used to deal with the incident.

To ensure that a debriefing exercise has a positive, constructive effect:

- It should be held as soon after the incident as possible.
- It should be multi-disciplinary and attended by the actual personnel involved in the incident. Services and functions that were not present at the incident may attend with the object of lesson sharing. It is preferable that the latter do not comment, contribute to the discussion or criticize the actions of those who were at the scene.
- Criticism and 'fingerpointing' should not be allowed. Debriefing should be a positive exercise that improves resources, expertise and performance in handling crises.
- Discussions should remain confidential.
- A summary of the learning experiences that result from the debriefing should be distributed to all functions that are part of the Crisis Management Team.

The communication function debriefing should review all aspects of the crisis communication procedures that were followed, how effective they were and what improvements may be necessary. Although these procedures and resources are situation and company specific, the assessment is likely to cover the following:

- the composition, functionality and accessibility of the Communications Core Crisis Group
- the availability of information about the situation and the ease with which the incident can be assessed, with a view to shaping communication strategy

- access to the Crisis Management Team and senior management for input on company objectives and strategy for handling the incident
- appropriateness and effectiveness of communications strategy and messages that are developed
- effectiveness of delivery
- adequacy of the communications centre and other resources
- effectiveness of feedback systems during the incident and post-crisis perception evaluation.

Information supplied by Honi Brian, Cormark Communication, cormark@ircradio.co.za

---

In summary, therefore, the key issues in stress management to consider are:

- Be aware of individual differences and select those who can thrive on a crisis.
- Be aware of how we react to crises and recognize the stress signals in ourselves and in others.
- Encourage positive coping.
- Prepare for extended pressure/stress periods.
- Keep communication clear, frequent and unambiguous to provide control and involvement.
- Consult and involve others as much as possible.
- Use relaxation and social support to counteract anxiety and panic.
- Expect the unexpected, and never underestimate the seriousness of the situation.
- Plan for crises – be proactive, not reactive.
- Review and learn from the process.

Criticism: The need for a broader perspective

The 'crisis management' approach aptly emphasises the need for crisis managers to be aware of and prepared to deal with their own stress and that of other stakeholders, including team members, during crises. It describes the attributes and skills that members need to function effectively in a crisis management team.

Problems arise, however, with the identification of these attributes as criteria for crisis team member selection. Even large corporations would be hard pressed to find a single manager, let alone a number of them, with all these qualities. The best one could realistically hope for is that a crisis management team, through good selection and judicious team-building training, would collectively demonstrate these attributes and skills.

Similarly, it is unfair to expect the average crisis manager, particularly in the heat of battle, to be much concerned with people's tensions, needs and feelings. Far more important is that the manager is secure in being a member of a tough, resilient, cohesive, reliable team of experts and that the supportive apparatus and services are in place to deal with casualties, the press, the general public and other potential distractions which merely add to stress.

The problems discussed above have been raised to illustrate that the crisis management perspective, as described, should not be followed too slavishly. A broader systematic approach to the control of stress in the crisis management team (and, in fact, the organization) is required.

A comprehensive stress management strategy could include:

- provision of counselling support for crisis management team members and other role players
- provision of trauma debriefing support for survivors and rescuers following life threatening disasters
- taking steps to remove or reduce unnecessary sources of occupational stress throughout the organization,
- provision of on-going stress management training for crisis management teams, in particular, and for employees generally.

To be truly effective, the strategy should be incorporated with the organization's crisis management plan and preparation. Most importantly, stress management training should form an integral part of team-building – the aim being to ensure that teams are resilient and tough.

Understanding stress

Crises are situations or periods of extreme pressure. Stress may be defined as *responses of an organization to changes in pressure or stimulation strong enough to disturb one's current state of equilibrium.* When experiencing this pressure it is the whole person as a psychological, social and biological being that reacts. The source of the pressure is known as the stressor. Pressures may arise in the external environment (e.g. being caught in a heavy traffic jam) or internally (e.g. concern about one's future prospects).

A working model of the stress response is shown below.

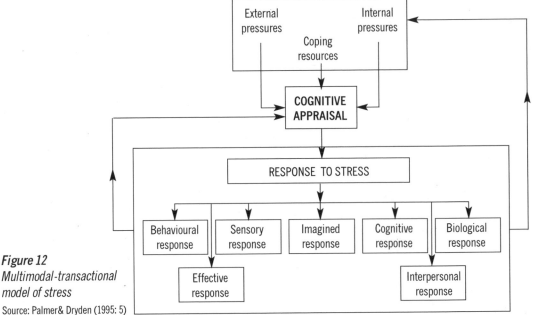

Figure 12
Multimodal-transactional model of stress
Source: Palmer& Dryden (1995: 5)

How one reacts to an event is determined by one's perception of it and one's perceived abilities to deal with it. In a prolonged difficult situation, one will repeatedly reappraise one's competence to cope successfully. The model distinguishes five stages of the stress reaction:

Stages 1 & 2 One experiences pressure emanating from an external or internal event and automatically appraises one's ability to cope with it. This appraisal is 'coloured' by one's history (e.g. cultural or family background, previous failures and successes in solving problems). If one decides that one can deal with or dismiss the situation, one's stress response is not likely to be activated (i.e. the pressure was not strong enough to disturb one's equilibrium). This appraisal occurs in an instant, it does not involve a long process of deliberation.

Stage 3 If one believes the situation cannot be coped with, one experiences a stress response. Psychological and physiological changes occur. Emotions like anxiety, guilt, anger are felt, and heart and breathing rates and blood pressure rise. During this phase, one will probably actively attempt, cognitively and behaviourally, to alter or control the situation, or escape from it and thereby reduce the pressure. This is the fight or flight stage. Some people actually 'freeze' with fear.

Stage 4 At this stage, the outcome of the action taken is reappraised. If one believes that one's attempts are not succeeding and that one is not coping, it adds to the pressure and strain in the situation.

Stage 5 Effective interventions by oneself successfully reduce or alter the external or internal pressures initially experienced and equilibrium is restored. People who have managed to cope with difficult life-events develop confidence in their coping skills and ability to handle future problems. They become 'tougher' and more resilient, their equilibrium is not as easily disturbed. On the other hand, people whose efforts are ineffective may experience prolonged stress and lose all confidence in their ability to cope, developing feelings of helplessness and inadequacy. Such stress reactions may continue just as powerfully for years, with serious consequences for the person's mental and physical well-being. For example, survivors of disasters may develop Post Traumatic Stress Disorders (PTSD) and suffer periodic, spontaneous unexpected 'flashbacks', reliving the trauma in all its horror throughout the rest of their lives.

Thus far, the discussion has centred around the damaging effects of stress. This is a one-sided picture. Below are some further facts about stress:

- Some stressful experiences are very pleasant, for instance, advancing through difficult competition to a sporting event final or winning the Lotto. Others, like disasters, are unpleasant. We cope with pleasant stressors more easily than with unpleasant ones. Pressures leading to moderate levels of stress are healthy. They activate constructive, problem-solving behaviour and peak performance which promotes feelings of well-being and growth. It does not matter whether the stressors evoking the moderate stress are pleasant or unpleasant.

- People differ in the amount of stress they can comfortably handle for peak performance. This is their zone of positive stress. Positive stress is sometimes referred to as eustress.
- Negative stress results from too much or too little pressure or stimulation. Negative stress is known as distress. Even too much pleasant stimulation can lead to distress, as can moderate pressure if it is prolonged.
- Stress accumulates. This leads to a gradual narrowing of our zones of positive stress, the amount of stress we can handle. Consequently, even moderate pressures can become too much to handle and cause distress. Most sufferers of stress related diseases have become so as a result of accumulated small stresses.

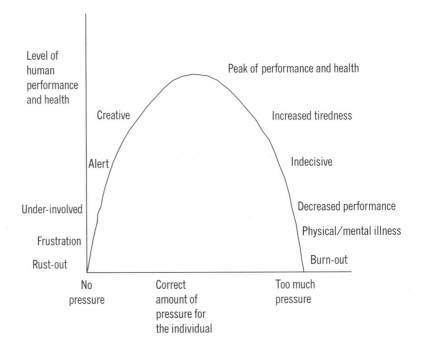

Figure 13
The relationship between pressure, eustress and distress.
Source: Palmer & Dryden (1995: 230)

Toward effective stress management

Human frailty is a major contributor in a high percentage of crises and disasters. Errors of judgement can turn routine situations into crises and crises into disasters. Errors of judgment are more likely to be made by people who find themselves experiencing undue pressure at the time.

Snapshot: Error of judgment

We had been cut off, separated from the rest of the Company, and trying for three days to find our way out of hostile territory. Too exhausted to carry on any further, we settled down for the night near a village. Then it started: hooters sounded, the enemy signalling to each other – imitating the cries of wild animals and birds – as they took up positions to cut off our retreat. This noise carried on spasmodically throughout the night to make sure that we got no rest. 'That's it! We'll have to make a frontal assault on the village and break out at first light,' we decided. Only Harry objected, but he soon fell in with our plan. We'd all agreed: 'No prisoners!'

We crept into position on the outskirts of the village and at first light, went in, guns blazing. There was no return fire, but a chorus of hooting, honking and other cries came from the geese, pigs and water buffalo as we disturbed their rest. The village was 'friendly', in our territory. It's scary to think of how close we had been to committing another 'My lai' massacre.

Comment by a Vietnam Veteran.

Not only individuals, but groups and organizations become 'stressed' with serious consequences for their health and productivity. Stress management, albeit included as part of a disaster or crisis management programme, should aim to:

- reduce the distress of victims (survivors), their families and any other people directly involved in and possibly traumatized by the disaster
- maintain the efficient teamwork and resilience of team members involved in containing the crisis
- build the mental and physical robustness and resilience of crisis management teams and their individual members
- reduce the incidence of distress in the organization and its workforce in order to enhance effectiveness and productivity and minimize the risk of stress induced errors of judgment
- timeously detect and neutralize the impact of stressors in the organization.

To be effective stress management is not simply attendance of a three-day workshop (they also have an important role), but a comprehensive, preferably company-wide, approach to the management and development of people. It should form an integral part of the company's crisis or disaster management strategy and should incorporate the collaborative efforts of medical, training, human resources and line departments as well as trade union representatives.

To realize these aims it is advisable that the Crisis Management Plan incorporates:

- structures for providing 'first aid' stress relief for traumatized victims as well as supportive counselling for team-members engaged in containing a disaster situation
- structures for regular 'refresher' training of crisis management team members – this would involve team-building and stress tolerance training under simulated, but pressurised conditions
- the development of multiple teams rather than a single team
- as a pre-emptive measure, involving the larger organization in stress management.

Snapshot: Mud-wrestling

Our first warning of the mudslide was a frantic 'walkie-talkie' call from our North gate security guard. I rushed to our assigned rallying point, where other members of our emergency management team were arriving. We waited, but our team manager was missing (later, we discovered that he was attending a Loss Control seminar in San Miguel, 500 km away). In the meantime, we were being given a running commentary by the guard. It seemed that the whole side of the hill was sliding down fairly slowly but relentlessly toward our plant. It had already penetrated our perimeter fence. Word had obviously got out because people were starting to stream out of buildings. We decided to do an on-site inspection as a team and then decide how to divert the mudslide to save the plant. We were hampered by people fleeing in panic in cars and on foot, but managed to reach the spot. Our guard had not exaggerated. One look was enough to convince us that we could do nothing to stop the slide. We hastened back to find that more than twenty people had been hurt, some seriously, in the panic. The path to the South gate (which was locked) was overgrown and too narrow for vehicles. A small truck had overturned injuring its occupants and a number of pedestrians.

The only heroes of the day were the guard, who stuck to his self-appointed task of keeping us accurately informed of the progress of the mud, and our clinic sister who, seeing the panic, called the local police (before our switchboard was abandoned), warning them to alert rescue services. When we arrived, she was calmly attending the injured and had cleared an area for rescue helicopters. In retrospect, if we had simply accepted our guard's report at face value, we could have informed the workforce and organized an orderly exit from the premises to a safer spot, before inspecting the slide. We were well trained for most emergencies, but who thinks of a mudslide. Strange! Because the area has an annual rainfall of over 2 500 mm and showers of 100 mm are not uncommon. My biggest lesson? There are a few:

1 Keep people informed and ensure that they trust what you say.
2 If there is a drama, someone will get wind of it and the grapevine will do the rest. The grapevine is faster than almost any official communication channel. It is also pretty accurate, but it adds to the drama and stress of the situation. As a manager, you need to have people's trust so that they

will give you the benefit of the doubt and not blindly believe the grapevine.

3 Ensure that people know what to do as first steps in an emergency. This can have a calming effect.

4 Let the clinic sister and security guard head the emergency team (a 'tongue in cheek' comment to draw attention to the value of employees commonly overlooked in the crisis and disaster management planning process).

Unpublished anecdotes and interviews on stress-related incidents, supplied by Clive Morgan

---

If stress is an identified culprit, indirectly or directly, in accidents, crises, and productivity problems, it makes good sense to tackle the problem at organizational, work-group or team, and individual job levels. Stressors, which exist in the greater organization, also have their impact on the Crisis Management Team and its members, individually. In essence, stress management involves three interrelated processes, viz:

- eliminating, neutralizing or containing harmful stressors
- reducing excessive stress or distress to more manageable levels
- building stress tolerance.

The role of managers in these processes is pivotal.

Hints for managers

1 Managers can alleviate some of the stress in the workplace by clarifying their expectations of employee's roles. Workers experience stress from not knowing what is required of them, where their responsibilities begin and end, having too much work to do in too little time, and so forth.

2 Managers should be aware that when subordinates experience excessive stress their decision-making ability and productivity decline.

3 Managers should be sensitive listeners when subordinates 'let off steam' and be prepared to help them reassess their situations to find constructive and productive solutions. Alternatively, they should guide them to seeking more professional counselling.

4 Managers should ensure that employees take sufficient leave annually.

5 Organizations should have policies and programmes that can help employees deal with their stress more effectively – managers should see that these policies and programmes are implemented and encourage subordinates to use them.

6   Managers should actively encourage the development of group cohesiveness, interpersonal networking and the general development of positive cooperative social relationships and mutual trust among employees. Promoting competitiveness tends to benefit a few at the expense of others, with a nett result of lower overall productivity than might otherwise have been attained.

7   Managers should promote open and frank discussion and airing of differences within their teams. Conflict should not be allowed to fester.

8   Managers should develop effective strategies for dealing with their own stress at work and be prepared, themselves, to make use of the stress management programmes that are in place.

9   Managers should be aware that they, themselves, may have difficulties in dealing with their own perfectionism, impatience, competitiveness and other personal traits which they, in turn, transmit to their subordinates, putting the latter under great pressure.

10 Managers should realize that productive, meaningfully employed employees enjoy higher job satisfaction and are more stress tolerant. The evidence indicates that managers should provide support and encouragement to help subordinates become productive rather than merely try to make them satisfied. Satisfied but unproductive employees are not necessarily more tolerant of stress.

Adapted from Arnold & Feldman (1986)

Managing job stress

All of us, from chief executives to cleaners, bring to work a residue of stress from our daily lives. In our jobs, we encounter further pressures which threaten our equilibrium. **Job stress** is the pressure we feel as we react to these threats. The threats may emanate from within ourselves (such as when we feel inadequate) or externally (such as dangerous work or a hostile boss). Excessive job stress has a destructive effect on our performance and decision making. It is also a serious health hazard.

Like individuals, organizations, and their subordinate departments and work groups, experience stress and can become distressed as a result of excessive pressures from internal and external sources. Modern organizations have to survive in a turbulent and often 'hostile' environment.

Effective crisis management requires early detection of signs and symptoms of stress within oneself, one's team and the organization as a whole. Early detection can lead to timeous preventive action or, at least, active preparation for a possible crisis. However our 'Stress Management' model advocates primary prevention. That is, crisis management teams and individual members should be engaged in continuous stress management and team-building training as part of their general preparation and readiness for handling crises. The model advocates further, that crisis managers should influence other managers and the greater organization to adopt active stress management strategies. The rewards for lowered job stress are fewer mistakes and accidents, better decision making, improved productivity and job satisfaction. Crucially, a major source of potential crises will be better contained.

Below are some warning signs of organizational, group (team) and individual job stress and possible remedial or preventive action.

Organizational stress

Stressed organizations become vulnerable under hostile external pressures. A single catastrophe could cause their collapse. Equally important, employees find the organizational climate in stressed organizations hostile and unpleasant. This adds to their job stress.

Warning signs of organizational stress include: increased and nastier political gamesmanship (particularly at senior levels), complaints about staff loyalty, scapegoating of middle and junior management, heightened union activity, increased disciplinary hearings, absenteeism and labour turnover (including resignations of experienced middle and senior managers), requests for early retirement or voluntary retrenchment, widespread insecurity and job dissatisfaction, and the filtering and censorship of information upwards and downwards in the organization. Relationships are noticeably guarded and untrusting.

Remedial action usually requires the immediate cleaning up of the internal social environment to re-establish a more congenial climate for work. Participative Organizational Development interventions involving joint problem-solving and goal setting relieve excessive pressure and help restore employees' sense of control of their jobs.

Pre-emptively, crisis managers (in conjunction with HR specialists) should monitor the organization climate and identify 'hot spots' so that appropriate corrective action can be taken.

Group stress

When two or more people join forces in largely face-to-face interaction to achieve a common purpose, we have the makings of a group. Teams are groups established to perform certain defined functions and achieve specific goals. Success requires sustained, committed, common effort by members who recognize their interdependence, are loyal to one another and identify with the team and its cause. As interpersonal loyalty grows, team solidarity or cohesiveness strengthens. This cohesiveness, with its common sense of loyalty, support and interpersonal bonding is a major source of strength for individual team members. It is particularly evident in crises. Strong teams endure long after their individual members in isolation would have succumbed to distress. This is why team-building is so important for crisis management teams. The development of a skilled, cohesive team takes time and constant practice.

Warning signs of stress in the team are when the team becomes over cohesive on the one hand, or starts losing cohesiveness on the other. Overcohesive teams represent tyranny to individual members, who dare not disagree with or question decisions for fear of being branded as 'disloyal', or a 'poor team man', or a 'maverick'. The pressure on the individual to conform and submerge his individuality eventually becomes personally distressing. Teams which cannot assess themselves critically and objectively are in serious danger of committing serious errors of judgment.

Teams which lack cohesiveness are teams in name only. Lack of member commitment is apparent when individuals regularly fail to attend meetings, fail to carry out assigned tasks, start sniping at each other, or have no constructive inputs to make during decision-making processes.

Remedial action in both cases is to look at the team and the pressures it is creating for its members, before apportioning blame to them as individuals. People should feel free to 'resign' from teams if they experience too much pressure from their membership. If necessary, disband the team and start again.

Individual job stress

Job stress arises directly out of the interaction between the person (who has hopes, fears, attitudes and skills), the work itself, interpersonal transactions (with peers, superiors, subordinates and clients) and environmental factors surrounding the job (organizational climate, group expectations, policies and procedures etc). Extraneous factors, like pressure to buy a new car, the state of the economy or the status of a person's job in the community, play an indirect but equally important part in that they influence one's attitude toward one's job.

A major source of stress is a person's role in the organization. A role is simply a set of expectations that other people in the organization (or team) have of one in one's job. An employee is expected to behave in a certain way. Stress arises when the employee experiences:

- role ambiguity – when the employee does not know what to do or what not to do. There may be uncertainty over job descriptions, authority to make decisions.
- role conflict – when the employee has to report to two bosses who make conflicting demands
- role overload – when the employee has too much work to do in the time available, or when the demands of the job exceed the employee's level of skill
- role underload – when the employee does not have enough work to do, or finds the work too easy and unchallenging.

Other stressors at work include such factors as noise levels, seating, lighting, temperature, interruptions, and so on.

Warning signs of excessive stress and potential distress include both physical signs and subjective feelings or symptoms. Among these are:

Physical signs
- increased alcohol use
- increased drug use
- increased tobacco use
- weight gain/loss
- increased/reduced activity
- increased/reduced sexual activity
- pacing the floor

- wringing hands
- worried look
- throwing objects
- kicking objects
- slamming objects
- changes in posture
- hyperventilation
- lack of control
- increased spending of money
- reckless behaviour
- loss of effectiveness at work

Symptoms
- palpitations
- chest pain
- unexplained sweats
- headache
- back pain
- tooth pain
- abdominal pain
- fatigue
- worry about being sick
- diarrhoea
- confusion
- tension
- aches and pains
- agitation
- repetitive thoughts
- increased interest in sex
- decreased interest in sex
- trouble breathing
- insomnia
- anxiety
- loss of appetite
- irritability
- neck pain
- thoughts or fears of failure

Adapted from Aronson & Mascia (1981) and Schafer (1987)

Job stress is an endemic feature of modern competitive society, even without the added pressures of the technological explosion, globalization and the often 'head-on' meeting of disparate cultures at the workplace. The potentially hazardous consequences of job stress cannot be swept under the carpet by any serious disaster or crisis management strategy. Of late, there have been increasing reports of disgruntled employees running amok, turning offices into killing fields and tragic incidents of 'road rage' over and above the problems of accidents, illness, absenteeism and generally lowered performance.

Remedial action involves active stress management strategies to be implemented by the organization, by managers and by employees themselves. Stress Management is an ongoing process of change involving the development of a healthy lifestyle. This is a process of change which must be monitored over a period because 'old habits die hard'. As part of improving and maintaining the health of the organizational climate, companies may consider introducing programmes involving cross-cultural interfacing, cooperative work training, team building, stress management, inter-personal skills training, effective competency-based job skill training, career-development and ongoing manager and supervisory skills development.

Conclusion

Stress, the reaction of the whole person, physically and psychologically, to stimulation or understimulation, is an inevitable part of life. Moderate doses of manageable stress are healthy, but excessive or prolonged, unrelieved stress leads to distress, which is a physical and mental health hazard. **Job stress**, which arises from various pressures at work, can lead to poor performance, ill health and dangerous mistakes, lapses in concentration, even irresponsible or violent behaviour. Not only individuals, but the teams and departments they work in and, in fact, entire organizations can become 'stressed'. Organizational stress destroys the work climate, creating a hostile, stressful, atmosphere for people to work in. Analyses of many avoidable disasters reveal the frequent presence of extreme pressure under which rational thinking and behaviour collapsed, leading to fatal mistakes. Crises and disasters are situations of extreme pressure.

Crisis management should incorporate provision for management of stress. Crisis managers must realize that they too are subject to the pressures of the crisis. Members may need debriefing or counselling to 'offload' the excessive stress they experience during a crisis. Planning and preparation of team members for crises should include stress management training and team-building in a continuous programme.

Finally, crisis management strategy faces the decision of whether:

- to take a conservative approach of confining itself to incorporating stress management interventions for team members only, or
- to take a broader, pre-emptive approach which influences the greater organization, through its managers, to accept and implement a policy of stress management involving all employees.

The discussion in this chapter favours the latter approach. Perhaps a conservative approach should be adopted initially, and once established, expand toward the broader perspective.

Sources

Arnold, HJ & Feldman, DC (1986) *Organizational Behaviour*, Singapore: McGraw-Hill.

Aronson, S & Mascia, MF (1981) *The Stress Management Workbook: an action plan for taking control of your life and health*, New York: Appleton Century Crofts.

Bland, M (1998) *Communicating out of crisis*, London: Macmillan Business.

Johnson, DW & Johnson, RT (1989) *Cooperation and Competition: Theory and Research*, Edina, Minnesota: Interactive Books.

Luthans, F (1989) *Organizational Behavior*, fifth edition, Singapore: McGraw-Hill.

Morgan, C, Counselling psychologist in private practice (2002) Conversation with the author.

Palmer, S & Dryden, W (1995) *Counselling for Stress Problems*, London: Sage.

'Pilot Error: Anatomies of Aircraft Accidents' in *Flying Magazine*, 1977, New York: Van Nostrand Reinhold (A Collection of twenty-five-accident reports, by the editors).

Schafer, W (1987) *Stress Management for Wellness*, Orlando, Florida: Holt, Rinehart & Winston.

Scott, MJ & Stradling, SG (2001) *Counselling for Post-Traumatic Stress Disorder*, second edition, London: Sage.

13

Summary and recommendations

In this chapter we provide a summary with recommendations of the major areas of concern in disaster management. They are drawn from a variety of sources which are acknowledged at the end of the chapter.

In planning for a crisis

1 Be prepared: develop a positive attitude towards crisis management, rather than wait for a crisis to hit you.
2 At all costs bring the organization's performance into line with public expectations. Remember the importance of corporate image and reputation and how a crisis could impact on these two pillars of corporate responsibility.
3 Be prepared to seize the opportunity during a crisis before your detractors. Above all, show your concern for the tragedy.
4 Appoint a crisis management team as a matter of urgency.
5 Identify potential crisis situations and devise policies for their prevention.
6 Formulate strategies and tactics for dealing with each potential crisis.
7 Appoint crisis control and risk audit teams to handle the emergencies.
8 Devise effective communication channels for minimizing damage to the organization's reputation.

9 Seek outside expert advice when drawing up crisis contingency plans. They may assist you during the crisis itself.

10 Put the plan in writing, but be flexible and be prepared to modify and adapt.

11 Look at what specialized training programmes may be required to ensure you have a professional group of people around you to assist in whatever emergency you face.

12 If possible test, test and test again.

In planning for communication in a crisis

1 Always be prepared to demonstrate human concern for what has happened. Don't appear callous and uncaring.

2 Be prepared to seize early initiatives by establishing the company rapidly as the single authoritative source of information about what has gone wrong and what steps the organization is taking to remedy the situation.

3 Have the names, addresses and telephone numbers of key media contacts available.

4 Identify other external groups, particularly pressure groups with whom communication will need to be maintained.

5 Research background information on any sections of the organization considered to be at risk; keep this up-to-date and available.

6 Set up a media room which can be used for media conferences and as a focal point for further dissemination of information. Staff it for 24 hours a day for as long as necessary.

7 Develop a cascade call-out list for all those designated and trained to facilitate communication during a crisis. Keep these lists updated.

8 Ensure that a senior public relations representative is part of the crisis management team, located in the emergency control centre.

9 Make sure the switchboard knows who to expect calls from, and to whom they should be routed, in the event of a crisis.

10 Keep a record of the number of calls received, their nature and general tone. As the crisis progresses, monitor the changes in attitude of callers.

In dealing with a crisis

1   In coping with a disaster, consider the worst possible scenario
    – and act accordingly.
2   At the outset of a crisis, quickly establish a 'war room', or
    emergency control centre, and staff it with senior personnel
    trained to fulfil specific roles designed to contain and manage
    the crisis. This may be on site or at a safe location close by.
3   Keep the media office up-to-date with developments as well as
    with the steps taken to control the situation.
4   Set up telephone hotlines to cope with the floods of additional
    incoming calls that will be received during the initial phase of
    a crisis. Have additional personnel trained to answer these
    calls.
5   Add credibility to your cause by inviting objective, authorit-
    ative bodies to help you end the crisis.
6   Be prepared to change plans as the crisis unfolds; and never
    underestimate the gravity of the situation.
7   Unorthodox methods of operation are often essential so the
    ability to bend the rules can prove vital. Have full authority to
    make snap decisions.
8   If you are at a branch office, communicate the situation to
    head office on a regular basis without dramatizing it.
    Remember that media reports which head office is receiving
    may tend to play up the most threatening aspects of the situ-
    ation and give a totally negative picture.
9   Staff must be prepared to cope with a high level of pressure
    and stress over an extended period.
10 When the dust has settled, look back to see what lessons you
    have learnt from the experience and modify your plans
    accordingly.

Communicating in a crisis

1   Start issuing background information about the organization
    as soon as possible after the onset of the crisis, demonstrating
    your preparedness to communicate during the crisis – while
    providing valuable breathing space for you to prepare accurate

media statements about what has gone wrong and what steps are being taken to remedy the situation.

2   If you are waiting for information or facts, do not fill the void with speculation or blatant untruths. Monitor what is being reported so that you can respond timeously.

3   Issue new media releases as more known facts become available. Make sure other personnel to whom the press may speak also receive copies of the releases so that 'everyone sings off the same hymn sheet'.

4   Announce the timing of press conferences as soon as possible to alleviate pressure from incoming calls. Prepare thoroughly for each media conference. Hold them on a regular basis at a pre-determined time.

5   Remember the media do not work from nine to five. Particularly if a major catastrophe is involved, the company will receive calls from all around the world, from journalists operating in different time zones. Staff the media office twenty-four hours a day, if necessary.

6   Develop a wide variety of information sources, cultivate journalists and local opinion formers. Keep up to date on reports on local radio, television and the press.
    Whenever possible, look for ways of using the media as part of your armoury for containing the effects of the crisis.

7   In communicating about crisis, avoid the use of jargon. Use language that shows you care about what has happened and that clearly demonstrates that you are trying to put matters right. If wrongly accused, go to any lengths to prove your case and get an apology.

8   Ensure that the organization has a list of responsible deeds and actions behind it to support the credibility of statements and claims made during the crisis situation.

9   Crises are an integral part of any organization's life and no business, regardless of its size or nature of operation, is immune to a crisis. Crises are not limited to national, headline-making disasters.

Corporate reputations can be badly damaged by mundane misfortunes like white-collar crime and strikes. But any unmanaged crisis can and will destroy the total infrastructure of any organization. A well-executed crisis plan will, however, limit long-term damage.

Sources

Cutlip, SM, Center, AH & Broom, GM (2000) *Effective Public Relations*, New Jersey: Prentice Hall.

Regester, M & Larkin, J (1998) *Risk Issues and Crisis Management*, second edition, London: Kogan Page.

Seitel, FP (2001) *The Practice of Public Relations*, New Jersey: Prentice Hall.

Skinner, JC, Von Essen, LM & Mersham, GM (2001) *The Handbook of Public Relations*, sixth edition, Cape Town: OUP.

Selected case studies

The Ellis Park soccer disaster

South Africa is eager to host a Soccer World Cup series and other international sporting events. The deaths of 43 soccer spectators at a soccer match in one of South Africa's premier sports stadiums in Johannesburg had the potential to end completely all the country's hopes of recognition as a possible international sporting venue. However, the effective handling of the crisis served to maintain a positive image of soccer and help shape popular opinion that, regardless of who was to blame, the authorities (the South African Football Association and the Professional Soccer League) acted responsibly in the way it handled the crisis.

What happened

On Wednesday, April 10, 2001, more than 65 000 fans went to a soccer derby at Ellis Park stadium in Johannesburg. Sadly, 43 were killed in a crowd stampede, and many more were injured. Television coverage showed in graphic detail the many injured and deceased being brought onto the sidelines of the field, and the difficulties faced by medical and security personnel in dealing with the crisis. What exactly happened that led up to the stampede was recently the subject of a judicial commission of inquiry.

More than 30 sports media were at the match. Each gave their own account of the events and apportioned blame in line with their understanding of the logistical responsibilities of each of the soccer officials.

Crushed by a human wave

Heartbreak, chaos and hope at the scene of South Africa's worst-ever soccer disaster

About 50 unidentified bodies of soccer fans are lined up as officials attempt to bring order amid chaos and sorrow at the Ellis Park Stadium last night.

Reports in local newspapers *The Star* and the *Sowetan*, on the day following the Ellis Park soccer disaster

Source: BlackRock Communications

Planning

The Crisis Centre went live within 24 hours of the incident. It was conveniently located at the Sunnyside Park Hotel, next to the offices of the Premier Soccer League in Johannesburg. It was equipped with:

- 6 networked computers with Internet and e-mail facilities
- 2 photocopiers
- 1 modem with ISDN line
- 10 Telkom lines
- 2 fax machines
- flipcharts and stationery

Roles and responsibilities

A team of more than 12 full-time staff worked in the crisis centre under a crisis centre co-ordinator, who managed five teams that were responsible for:

- victim liaison
- strategic events
- government co-ordination
- internal communications
- media liaison

Strategy

An accommodative strategy was decided on in which the truth would be told completely, guilt acknowledged, responsibility accepted where appropriate, and corrective action, such as helping victims, would be taken.

Due to the legal implications of the crisis for a number of stakeholders, a public apology could not be made. Because South Africans do not respond well to people who do not apologize, a generic key message was devised that included the word 'sorry' – This was the message:

> 'We planned thoroughly for the event, yet something went wrong and we are sorry that something went wrong. We must find out what it was so that it never happens again.'

The media allegations were numerous – the sale of too many tickets, teargas used to control the crowds, moratoriums disobeyed by playing the game at that venue, and more. To try to respond to any of these without having all the facts, would have been disastrous. It would have meant defending each allegation in the media one by one as new allegations surfaced almost daily.

It was realized that due to the Easter public holidays it would be impossible to get messages out in the first three days following the disaster, so all energies were focused on notching up as many 'action taken' stripes as could be generated over the Easter weekend. These included:

- hosting a media briefing within 24 hours to deliver a central message of empathy in a controlled environment in which the media could not play one soccer party up against the other
- establishing the Crisis Centre
- holding a wreath-laying and memorial ceremony
- setting up a 24-hour hotline
- offering free trauma counselling
- visiting families of the deceased and victims in hospital
- establishing a relief fund

In this way, it was hoped to shift the focus away from 'naming and blaming' to what was being done.

Since people interpret information according to what affirms their existing opinions and beliefs, five key crisis response actions were used to show evidence of doing the right thing, while acting on popular opinions and beliefs. These were:

- acting extremely fast – because people believe that responsive people are prepared and organized and therefore good
- acting responsibly – because responsible people/organizations are less likely to make mistakes
- showing empathy and concern – because people who care are usually good people
- maintain the unity of top management as a team – because a unified team has nothing to hide and therefore individuals have no reason to cover their own backs.

Defining clear and measurable objectives

A set of objectives was agreed upon that could be measured at the end of the crisis:
- to minimize mis- and dis-information in the media
- to manage the expectations of victims and families of the deceased
- to protect the reputation of Ellis Park as one of South Africa's premier sporting facilities
- to maintain and protect the relationships that underpin soccer, the brand and the values it embodies for sponsors, financial stakeholders and fans
- to minimize the impact that the crisis could have on South Africa as a tourism and investment destination

In order to achieve these objectives, there were a number of critical success factors that needed to be taken into account in the strategy:
- a multi-sectoral approach, which included various government, business and soccer officials seen to be working together
- changing the situation from being driven by the media to being proactive and in the driving seat of message management
- minimizing rumour and speculation by establishing channels of communications to put forward the facts directly to key stakeholders.

Due to the importance of message management during a crisis, a number of key messages were agreed upon:
- The primary focus is on the families of the deceased and the injured.
- Support the President's call for a Judicial Commission of Inquiry.
- The Crisis Centre is working round the clock to get the victims identified and offer them support.
- The loss of life is regretted and actions will be taken to ensure that this never happens again.
- It is a national pain to lose life at a sporting event.
- Thousands of letters, faxes and calls from around the world, offering South Africa condolences and support, have been received.

Execution

In order for the South African Football Association (SAFA)/ Professional Soccer League (PSL) Crisis Committee to be seen to be 'doing the right thing' and to shape positive public opinion towards the handling of the crisis, the following actions were taken:

- financial compensation for families and relatives of victims
- establishment of a 24-hour toll-free hotline
- establishment of a relief fund
- a wreath-laying and ritual cleansing ceremony
- a memorial service
- post-traumatic stress counselling

The question was posed 'What would it take for people to say that soccer officials acted responsibly following the crisis?' The answers were:

- Establish a Crisis Centre from where message control and knowledge management could be implemented.
- Host a dignified and formal ceremony that showed order and control within a soccer stadium.
- Hold regular media updates and briefings to show the media that the organization was transparent.
- Humanise the face of the Crisis Committee with a senior soccer official who could make credible media, business and government presentations.
- Do not be emotional, but act on evidence and support responses with facts and figures.

Evaluation

The success of the Crisis Management and Response to Ellis Park by the SAFA/PSL Crisis Committee was judged by the following criteria:

- whether people continue to go to Ellis Park (Yes)
- whether people continue to go to Kaizer Chiefs and Orlando Pirates soccer matches (Yes)
- whether the families of injured and deceased received satisfactory service (Yes)
- the total contributions made to the relief fund (R2,3-million)

- whether the soccer fraternity and their sponsors have positive perceptions about the handling of the crisis (Yes)
- the number of balanced editorial articles (70 per cent)

This case study was adapted from a presentation by BlackRock Communications which won a Silver Award at the 2002 FM/PRISA/PRISM competition.

The Nairobi bomb blast at the American Embassy – a diagnosis of failure

Friday, August 7, 1999 was like any other ordinary day in Nairobi; the Friday rush, the traffic jams and crowds milling in streets going about their business. The American Embassy located at the junction of Moi and Haille Selasie Avenues was busy as usual. Nothing seemed out of character. The nearby bus stage and office buildings had endless streams of people coming in or going out. This day, however, was going to be devastatingly different.

What happened

At 10:30 a.m., an explosion sent a fireball and an ominous cloud over the city's skyline. A training editor of one newspaper, Frank Whalley, gave this account:

> '*I knew at once it was a bomb. I had heard that awful sound before.*
>
> *First the explosion – louder than anything you can imagine – then a series of smaller secondary blasts as most buildings fold in on themselves, and the shock waves echo on, destroying whatever they hit; shattering windows and bringing whole sections of buildings shuddering to the ground.*
>
> *And then the terrible silence; Kenya was under attack. Then the situation melted into chaos, despair and hopelessness.*'
>
> Sorce: *Daily Nation*, 8 August 1998

The American Embassy was the target of a terrorist attack. Details emerged later. The intention was to drive an open pick-up into the basement of the building, loaded with a powerful explosive, and then detonate the bomb. According to experts, given the reinforced concrete structure of the building, this would have caused the building to be uprooted and with it, other buildings within a half kilometer radius. The buildings would have come tumbling down like a pack of cards. An eye witness said that a yellow open pick-up van raced down the Kenya Railways round-about and reversed into the American Embassy building. Two men jumped out firing machine guns at the Embassy Security staff. Their plan to enter the basement was, however, thwarted. The bomb was detonated and the entire city was thrown into shock and chaos.

Planning

At the scene, analysts painted a picture of a complete lack of planning for such an event. Gitau Waroi, a columnist with the *Sunday Nation*, reflected:

> '*When the Police and General Service Unit (GSU) (a paramilitary service) and the army arrived in numbers, which was long after the blast, they came with guns and in full battle-dress as if they were coming to confront some major public disorder, like a riot; none of them had a first-aid kit – yet even when they arrived, they had little to do other than keep pushing the crowds to remain at bay.*
>
> *At the scene, the situation was chaotic and completely disorganized. The public had taken matters into their hands, fetching the dead or injured and using private vehicles to take casualties to the hospital. Private construction companies moved earth-moving equipment to the blast site. They were ready to help in whatever way they could. Coordination was, however, lacking. Meanwhile, efforts to take the injured to hospital continued. Thousands of Kenyans scrambled to the scene and tried to rescue the living and remove the dead from the wreckage.*'

Source: *Sunday Nation*, 16 August 1998

Strategy

Within an hour, hundreds of Kenyans were organizing aid to hospitals to cope with the disaster. By the evening, Kenyans were still queuing in hospitals to donate blood. 'I had to wait for three hours to donate blood,' Kabando wa Kabando, a Policy Researcher, was quoted as saying.

'We had to literally start turning people away,' said Michael Sheldon, Administrator of Nairobi Hospital. This shows the quick response of people to help by donating blood. Other Kenyans cooked food and took it to the hospitals and the blast site to feed relatives and rescue workers. Blankets and medical supplies were rushed to hospitals by private and civic organizations.

In the hospitals, the doctors, nurses and every cadre of staff were summoned for emergency services. Various types of medical specialists were flown in. A visit to hospitals painted a painful experience. Kipkoech Tarus of Nation Newspaper recounts, 'A tour of the eye-unit and eye-ward revealed a miserable picture of patients with deep facial cuts, severely perforated eye sockets and swollen faces.'

Tactics

It was evident the authorities were caught napping by this catastrophe. There was no coordinated press briefing, until foreign rescue missions provided briefings.

There were four rescue teams from Israel, France, Kenya and America. Although they came long after the blast, some as late as twenty-four hours, they made a significant difference. The Americans were initially accused of being interested only in securing their documents and searching for American Embassy staff. However, the Israelis won the admiration of Kenyans as they arrived with a variety of tools and, most important of all, sniffer dogs. These teams hardly slept for six days.

Meanwhile, a disaster fund was established to coordinate fundraising efforts to cope with the disaster. The members of the disaster fund agreed to work as volunteers. PricewaterhouseCoopers agreed to offer free audit services. The committee had offices in the Uhuru gardens with notice boards containing information on the dead and injured.

Lessons to be learnt

Valuable lessons have emerged from the bomb blast experience. Obviously, Kenya lacks a national disaster response mechanism. The ambulance services were uncoordinated as a result of a lack of a central crisis communications centre. Though there were City Council fire brigade vehicles on site within an hour, it was clear they were not sufficient; they could not cope with their short ladders and tanks that ran out of water. Nairobi has no plan for water points for emergencies.

As people under the collapsed Ufundi building lay under the rubble, there was no clear command for the rescuers till hours later. Senior government people kept on streaming to the site after being summoned over National Radio. This worsened the situation as useful time was spent briefing them on their arrival.

The ordinary Kenyan people, however, responded in ways that were effective. They gave what they could: blood, food and labour. The medical fraternity, though overstretched, worked extra hours to cope with the disaster.

After initial denials of a security lapse, the United States of America later admitted this, but somewhat belatedly. US Secretary of State Madeline Albright was quoted as saying, 'Reports suggest that responsibility for this failure must be shared equally and I accept this' but generally America was not seen as doing enough to lessen the suffering of Kenyans.

The rescue effort was called off after six days. The final toll was 254 dead, hundreds in hospital, and five thousand injured. Non-governmental organizations and counselling professionals were contracted to help those affected by the disaster. There was bickering over finance, particularly over alleged discrimination against local professionals by donors keen to fund foreign experts.

The media relations response was also totally inadequate. There was no clear communication strategy. However, as foreign teams arrived and started media briefings, the government changed tack. The Kenyan authorities appointed Major-General Agoi to coordinate media relations efforts and, although official press briefings were now accessible, it was generally described as 'too little, too late'.

Mr Peter Kiuluku, Public Relations and Marketing Manager of ESAMI (Eastern and Southern African Management Institution), provided valuable background information for this case study.

Community relations: Shell Oil's response to crisis in Nigeria

Background

Shell Petroleum Development Company of Nigeria has operated in Nigeria for almost 40 years, involved in oil and gas exploration, oil production and marketing. In 1958, Shell struck oil in Ogoniland, which set in motion a process that dramatically affected not only Ogoni society, but Nigeria as a whole. Shell is the first, and largest, of the six major oil-producing companies in Nigeria. Today, oil accounts for over 90 per cent of Nigeria's export earnings and some 80 per cent of government revenue, controlling the entire Nigerian economy. Shell is responsible for about half of the country's entire oil production.

Shell was the exploratory company to gain first access to onshore wells in the Niger Delta region of Nigeria, the source of over 90 per cent of Nigeria's oil. A joint venture agreement with the government-owned Nigerian National Petroleum Corporation (NNPC) gave it a substantial advantage over smaller oil companies like the French-owned ELF and Italian-owned AGIP.

Shell also has a 25 per cent interest in the Nigerian Liquefied Natural Gas (NLNG) Project. The Nigerian government's stake in the project is 49 per cent.

Shell employed 4 700 Nigerians full time and had a contract workforce of 20 000. Given its powerful commercial position, its actions and decisions affect the livelihood of ordinary Nigerians at the gas pumps, as well as in the communities in which Shell installations are located.

Chronology

For the Ogoni, who live in this region, the environmental and social costs of oil exploitation were high. In 1990, the Ogoni organized themselves in the Movement for the Survival of Ogoni People (MOSOP). The President of MOSOP was the well-known Nigerian writer and poet Ken Saro-Wiwa. The Ogoni Bill of Rights (OBR) outlined their demands for environmental, social and economic justice. The OBR opposed the revenue allocation formula under which the federal, state and local governments have almost complete power over the distribution of oil revenues. The Ogoni felt they were not adequately compensated for the take-over of their land by the oil companies and the environmental damages they suffered.

In Umuechem on November 1, 1990, youths attacked a Shell rig and flow station in a tiny community in Rivers State. Armed with machetes and guns, they demanded compensation for destroyed farmlands. Citing his employees' safety, Shell's Eastern Division Manager sought police protection. A mobile unit responded and three officers died in the ensuing confrontation. Police reinforcements, according to Amnesty International, shot and killed about 80 villagers and destroyed over 400 homes. Shell reported that although many people lost their lives, only seven were identified. This massacre gave impetus to the creation of MOSOP.

Armed men attacked Shell's rig camp and flow station at Ahia on March 7, 1992. They held rig workers hostage and destroyed or looted property worth thousands of dollars. They demanded compensation from oil revenues and the building of new roads in the community. Shell again turned to government to quell the disturbances and shut down production for nearly a month.

Led by the charismatic Ken Saro-Wiwa, a rally was organized in Ogoniland in January 1993 where more than 300 000 Ogonis protested against Shell. Escalating tensions erupted in violent attacks on Shell staff and property. Shell recalled its staff.

In 1993, Shell withdrew from Ogoniland, because of the hostile attitude of the local community to the company's activities. Later that year, Shell went back under the protection of the Nigerian military. The peaceful protests of the Ogoni against the building of yet another pipeline were answered with gunfire. Several conflicts occurred between July 1993 and April 1994 between the Ogoni and their neighbours, the Andonis. Critics charged that available evidence, including the sophisticated weaponry used by the Andonis, indicate that governmental authorities were probably behind the supposedly ethnic conflicts. MOSOP blamed the government and Shell. Meanwhile, a worldwide campaign against Shell led by organizations such as Green Peace and Amnesty International exploded in the international media.

A crisis erupted in Nembe Creek on December 4, 1993, when a

group of armed youths attacked the Nembe production staff, vandalizing and looting Shell property. They demanded Shell provide electricity, build roads, and hire local labour. Shell again turned to the government, which sent in a Mobile Police unit, and shut down production for five days.

In Ogoniland, May 1994, four Ogoni leaders, all former members of MOSOP who allegedly had criticized Saro-Wiwa for losing sight of the original purpose of MOSOP, were murdered during a public meeting.

The aftermath of the 1994 murders was the arrest, conviction and hanging of Saro-Wiwa and eight others. The international outcry resulting from the execution was directed at both Shell and the government. Shell was blamed for quietly sanctioning the hangings. It denied both complicity in the executions and a role in environmental pollution. Subsequently, Shell felt compelled to stop its denials and embarked on a vigorous campaign to restructure its badly tarnished image. In March 1996, non-partisan elections were held in Nigeria to fill all local government seats with elected civilian Chairs and Councillors. It was widely reported that all persons associated with MOSOP were prevented from participating, and those who tried to were beaten or detained and later disqualified. Since the election of President Obasanjo, the situation of the Ogoni people has reportedly improved.

Goals and objectives

As the crisis unfolded, the company worked with some of its publics to keep its operations functioning. One of the publics, state and federal governments, provided tactical support to enable Shell to control vandalism at its installations. Another public, Shell employees, was used as goodwill ambassadors to the local communities.

The mass media, consisting of the Nigerian and international media, published and broadcast reports about the Shell crisis to the world.

The Nigerian media, excepting government-controlled radio and television stations, have always been vocal critics of the government. Investigative reporters went to great lengths to ferret out evidence of Shell's collusion with the government in the crises at its oil installations. The world press was even more vocal. Coverage by the world press focused more on Shell's human rights record than environmental damage.

The community in Shell operating zones went from being latent to very active. Nigerian ex-patriots protested at Shell offices worldwide and wrote petitions and editorials calling for the boycott of Shell products. International human rights and environmental groups were highly active. Their actions, including protests at Shell headquarters in London, scathing editorials in the world press and on the Internet, and lobbying activities, drew further attention to the Shell crisis.

Shell's impact objectives (planned outcomes):

1 To establish/re-establish better dialogue with the local oil producing communities in order to stem further problems
2 To use opinion leaders/community leaders as a bridge for dialogue
3 To engage in continuous and open two-way communication
4 To boost spending on both environmental and social or community projects:
   Environmental
   a Replace and bury aging oilflow pipelines to eliminate leaks that pollute and kill marine life
   b Reduce the level of greenhouse gases caused by open gas flaring
   Social/Community
   a Build community clinics, schools, and youth/job training centres
   b Provide scholarships and jobs for indigenes of the oil-producing areas
5 To lobby the federal government to introduce economic reform, including privatization, in order to ease the economic despair of the affected communities

Shell's process objectives (information dissemination):

1 To gain positive international media coverage of its efforts to boost social and environmental spending in the oil-producing

communities through the use of international media, speeches and interviews and press conferences, and through Shell's website and the company's special reports and bulletins

2 To gain positive national media coverage of its efforts to re-establish dialogue with the community through the use of press releases, speeches and interviews, and Shell progress reports

Strategic tactics and techniques

Shell adopted several effective messages to meet its objectives. It promoted itself as a company that tries to balance business with active citizenship and work in partnership on community activities; whose business activities and decisions do not deliberately infringe on the community's basic human rights or affect their environment; and as a law-abiding, non-political and non-sectarian company that listens to its publics.

Results achieved

March to October 1996

Shell provided the media with regular press releases on its efforts in the communities. Headlines such as 'Shell Rebuilds Hospital', 'Shell Joint Venture To Reduce Gas Flaring In Nigeria', 'Shell Helps Youth Find Jobs', and 'Shell Helps Ogoni Hospital' helped the company to show the public how its relationship with the communities had evolved.

Shell also used press releases and its website to respond rapidly to its critics even as it worked to improve its image.

Shell took journalists on tours of some its community projects in the Niger Delta.

Shell sent its public relations team on a whirlwind tour of major European and North American cities to tell 'their own side' of the story. Shell executives were interviewed on Cable News Network (CNN), the British Broadcasting Corporation (BBC), and the Canadian Broadcasting Corporation (CBC), among others.

Shell's community relations officers, sometimes accompanied by top officials, had 23 000 meetings with opinion leaders in 1996, compared to 1 300 meetings in 1995.

Shell's top executives met with influential Nigerian organizations, such as the Rivers State Foundation (RSF), a U.S.-based non-profit organization whose members come from the oil-producing communities.

Shell's 1996 Community Report provided detailed information on Shell's contribution to the oil-producing communities in healthcare, vocational training, education, agriculture and community achievements.

February to July 1997

- Shell initiated a youth training programme in Ogoniland, which also received medical benefits. Other communities benefited from scholarships and potable water schemes.
- Shell's Group Managing Director, Phil Watts, spoke at the 4th African–American Summit in Harare, Zimbabwe, on how multinationals can help in 'economic development, human development and environmental sustainability'.

Conclusions

Benoit (1994) contends that when our image or reputation is threatened, we are motivated to explain, defend, justify, rationalize, apologize or make excuses for our behavior. The same principles apply to organizations. Shell's attempt to rebuild its image falls into broad categories of image restoration strategies discussed by Benoit: denial, evasion of responsibility, reduction of offensiveness, corrective action, and modification.

At the onset of the crisis, Shell denied that environmental pollution in the oil-producing areas was related to its activities. The company also denied that it had any influence with the military government; therefore it could not stop Ken Saro-Wiwa's execution. The company argued that, out of concern, it used 'quiet diplomacy' to try to get the Saro-Wiwa death sentence commuted.

Shell evaded responsibility by 'scapegoating'. The company argued that charges of environmental devastation were distorted. It blamed local thugs for sabotaging Shell operations to gain attention, and finally, indirectly blamed the government for shirking its responsibilities to its oil-producing communities.

Shell focused on its pioneer status in the Nigerian oil industry and its enormous contribution to the economy. It contended 'we are the largest oil company in Nigeria and we make the largest single contribution to the country and are a force for good in Nigeria', and pointed to its use of local subcontractors and the technical training of staff members.

In Shell's use of corrective action, the company promised to rectify the problem and promised to act to prevent a recurrence. Shell went on to identify specific steps to repair its damaged relationship with the communities, including restoration of the environment, increased spending on community and social projects, and open and improved communication with the oil-producing communities and other publics.

Although Shell had initially tried to take a nonpolitical stance, Shell's general manager, Brian Anderson, used his considerable influence and high profile to lobby Nigeria's military regime to initiate economic reforms. Anderson believes that economic reforms, including privatization, and the government's reduced stake in the joint venture with Shell will lead to an improved standard of living for members of the oil producing communities in particular, and the nation at large.

Shell's relationship with the communities continues to evolve. The company's efforts to foster better relationships with the oil-producing communities are still evolving. Dialogue and some openness are gradually replacing the mistrust between the two.

Source: VanSlyke Turk, J & Scanlan, LH (1999) *Fifteen case studies in international public relations*, Florida: The Institute for Public Relations, University of Florida

Brand crisis management

Recently one of the world's biggest website portals, Yahoo!, managed to get its name mentioned in most media around the world and it wasn't positive news that brought public attention to the Yahoo! brand.

The stimulus was from Yahoo! in France, where its online auction site included Nazi paraphernalia – Nazi medals, clothing, ephemera, and other artifacts – all available to the highest bidder. The crisis was over the French government's ban on racist representation in all media. The French authorities saw Yahoo!'s auction listings as an infringement of its jurisdictional authority.

This was not a predictable crisis, even for the most experienced brand experts. But it happened to ensnare the well-respected

Yahoo! brand, which had to suddenly deal with a problem fraught with negative associations.

Dot-coms haven't been around long enough to have gathered lots of crisis management experience. But now, it seems, the honeymoon period is over. Everyday life has at last caught up with the Internet which recently seemed to offer its dot-com residents some measure of protection from critical analysis. But admiration for the online phenomenon has given way to realistic appraisal of dot-coms as business entities.

Established offline brands around the world have been in crisis situations hundreds of times before and, over time, they've developed crisis management programmes. A good example of this is the Australian biscuit brand Arnott's.

Some years ago, Arnott's, an icon of Australian identity as resonant for Australians as the famous Vegemite, was faced with an extortion threat by a criminal who claimed to have poisoned some of the company's product. Experts advised Arnott's that such a threat was indeed capable of being carried out swiftly, the extortionist having given the company just three days to respond to his demands.

What would you do in Arnott's' shoes? Of course, the company was prepared for this unfortunate eventuality. It recalled all its biscuits, destroyed them, produced a new and totally different package design and, within days, was ready to relaunch the brand Australians had known for generations.

In the meantime, the PR department spun a story designed to appeal to the Australian consumer's sense of loyalty and, almost, patriotism: the danger of Australian companies being lost to overseas interests and the undesirability of a well-loved Australian company going under at the hands of international competition. The strategy worked well, garnered plenty of community sympathy and support, and prepared the way for Arnott's hugely successful re-launch, all of this managed in 72 hours.

Brand crisis management programmes predict possibilities and prepare for hypothetical eventualities. Just like regular fire drills, they set out step-by-step instructions for all players. Plans are developed over several years and tested to minimize the ill effects of crises should they occur.

Adapted from Lindstrom, M (2000) *Brand Crisis Management*, Bernstein Communications [Online] Available: www.bernsteincom.com/nl/crisismgr001015.html

Managing the threat of a cyber crisis

The Internet is a double-edged sword. Because of its reach, speed and interactivity it can destroy an organization's image and credibility or be a useful tool to manage a company's reputation, and help it to respond in crises.

Today, the news of a disaster can be spread globally almost instantaneously via the Internet. What do you do when the story contains elements of untruth and is damagingly hurtful to the business of the company? Local issues can very easily become a global issue. The Internet today enables activists and other support groups to mobilize worldwide support for a cause.

The new media can radically alter the perception of your company. In the past, it used to be the conventional news media that created that perception, but now a single message on the right Internet discussion board can do damage to a company's hard-earned reputation.

The reason the web can make or break your organization is because it is largely unmediated – the gatekeepers of the traditional media are often absent – and because such messages are copied and replicated often with scant regard for accuracy. The Internet could be described as the repository of conspiracy theories and falsehoods that travel like a veldfire. Because of the Internet, issues spread more quickly and widely now that anyone can publish on the web.

This means that local issues can now quickly become global issues and that these issues can now spread to mainstream media like a virus. For instance, the Bill Clinton–Monica Lewinsky affair first broke on a web site. The Internet has become a primary news source for journalists and organizational web sites are now used widely by journalists.

Increasingly, the Internet has meant that people receive information on the run and that the opportunities to prevent bad messages spreading have diminished. It means therefore that there will be little available turnaround time in a crisis and that immediate action will be necessary to protect your organization's hard-earned reputation.

By definition, a crisis is only one when it becomes public. Before that, it is a simple business problem. But, as in all crisis management, the secret is not to have one. So the Internet is as much about crisis avoidance as crisis management.

Some of the effective reputational defence responses on the Internet include:

Updating the organizational website

One of the key sources of information for journalists is an organization's website. They will often visit the website to gain background information on the organization, often prior to speaking directly to a source. Many sites are designed to incorporate the latest media releases, and even if they are not, it is a good idea to carry full information about the crisis in the site. The solution is so simple: post your crisis information timeously, clearly and prominently.

This means, of course, that the webmaster must be part of the strategic crisis communication plan and important information must be made available to him/her for uploading as a matter of urgency.

Monitoring of the Internet

Ongoing monitoring of the Internet becomes essential to identify hostile messages, and hate sites early so that you can correct misperceptions and false statements before they do damage. It is therefore important for the corporate communicator to participate in newsgroups, forums, bulletin boards, and so on to correct information that has been misquoted, wrongly excerpted or fabricated. Watch key sources of information, rumour and discussion about your organization, its products, services, stakeholders and employees, and track reporters and others who use these sites for stories.

Establish direct communication

It is important to realize that electronic communications can allow you the opportunity to speak directly to your target audiences with no media filtering. This means that you will have to revisit your use of various technologies such as e-mail, SMS and group bulk e-mailers.

One of the major mobile applications that can be employed by an organization in crisis is the Short Message Service (SMS). SMS is the most rapidly growing mobile phone technology worldwide, which enables companies to communicate with staff, and, in fact, all stakeholders individually or on a group basis. Because it is an instant and private means of communication it can be used to alert stakeholders about a crisis.

Crisis communications can be sent via WebSMS, which delivers bulk SMS via the Internet. E-mail SMS means corporates can leverage the current e-mail infrastructure to deliver SMS to communicate primarily with people in the company's e-mail address book. FTP SMS uses the file transfer protocol to send large volumes of SMS messages to cellphones. It can send messages to hundreds or thousands of recipients. These are powerful tools that can be enabled in times of organizational crisis.

In the US, a recent annual survey of companies by the Computer Security Institute and the FBI revealed that 90 per cent of all firms have had some type of IT security breach in the past year. Ironically, the FBI itself has fallen victim to hackers.

Every day, hundreds of new web pages are listed on one popular hacker's web sites after being defaced. In some cases, the culprits are juveniles posting pornographic material and admonishing the system administrators to implement better security. In others, political or environmental hacker activists are protesting the targeted companies' business practices or products.

Website defacements are only one – relatively benign – form of computer intrusion. Dozens of other types of computer-related attacks and security breaches occur every day, including theft of customer credit card information, theft of intellectual property, extortion and threats of violence. And these are the attacks in which the motive is clear. In many other cases, the targeted company can only detect that an intrusion occurred, without being able to determine either the reason for the attack or its precise effects.

Here are a few questions that executives should consider:

- What would you do if hackers brought down your web server for a day?
- How about five days?
- What if they used your systems to launch 'Denial-of-Service' attacks against other companies?
- What would you do if an anonymous investor published false financial information about your company on the Internet?
- What would you do if confidential information was being e-mailed out of your network?

- Are you sure you would even know that it had occurred?
- If you received a threat to destroy your network unless you paid R100 000 for security consulting, as a company in South Africa was, what would you do?
- Could you prevent the threat from being carried out?
- Are these threats even credible?
- Do you fully understand what the costs or impact to your business would be?
- Who would you contact? Law enforcement?
- Could you handle this in-house? Should you?
- What if the source appeared to be from another country?

Information Technology: a complex and sometimes hostile environment

Organizations often do not understand the risks to their information assets. A thorough understanding of the risks to an organization's information infrastructure is crucial both to management's ability to make decisions on where controls should be implemented, and to its ability to manage crises successfully. Few organizations invest in a proper risk assessment before implementing controls. Even fewer understand and qualify specific threats in order to evaluate risks accurately. The consequences can be profound. Not only are some threats overlooked, but also resources and budgets are misapplied to threats that do not exist or have minimal impact.

Computer and telecommunications networks are fostering a revolution in the way organizations do business. The unprecedented ability to interconnect every aspect of a company's business through the use of networks provides myriad opportunities for increased efficiency and enhanced communications. The emergence of e-business is fundamentally altering both the way companies function internally and the way they interact with suppliers, partners, customers and governmental agencies. These changes are international, creating new markets and competitors.

With this unprecedented connectivity come new risks for the information infrastructure upon which organizations are building their e-businesses. The electronic assets of many of today's top companies are inseparable from their physical 'bricks and mortar'. In this environment, the combination of global connectivity, employee mobility and rapid technological change creates new opportunities for fraud, theft, extortion, pirating, industrial espionage and business interruption.

Technology is not the only source of risk to information infrastructures. Political, physical, environmental, legal and regulatory issues are all factors contributing to the creation of a multi-dimensional problem.

While prevention is the preferred course of action, no security measures are perfect. Organizations must be prepared to quickly detect and effectively respond to the threats they face in the ever-changing e-business environment.

Insiders or outsiders?

So where do these threats come from? Traditional wisdom holds that insiders are the greatest threat to an organization. This is based on two assumptions: first, that insiders have access and second, that they have knowledge of a company's systems, applications and processes.

However, the Internet and e-business are creating a new environment. Consider these facts:

1 Companies are connecting to the Internet as quickly as possible. These connections occur with little planning and few controls, creating a whole new level of access from the outside.

2 Because of electronic supply chain management, B2B e-commerce and customer relationship management (the electronic connection of companies' businesses to their suppliers, customers and partners), the traditional boundaries are becoming blurred. For example, a subcontractor hired by one of your suppliers (without a background check and little management supervision) may now have access to and knowledge of some or all of your business applications and systems.

3 Most companies no longer build their own proprietary business applications; instead, they purchase standard, off-the-shelf applications for such things as finance, customer relationship management and order management systems. This standardization allows outsiders to use applications without detailed internal information. These and other factors have altered the threats companies now face. The distinction between an outsider and an insider is decreasing rapidly.

What should you be doing?

The first step any organization's management should take is to make an effort to clearly understand the threats to its information infrastructure. When determining these threats, executives need to look beyond simple technical vulnerabilities in their company's computer systems and networks and consider the full range of threats, including those that involve people and issues, operational and business processes, culture, physical issues and legal and regulatory concerns.

Additionally, management needs to understand that, while technology enhances capabilities by providing access, people, whether trusted insiders or outsiders, are the source of many threats. It is vital to understand the source and motives behind illicit activity and the targeting of specific businesses or industries by individuals or groups. With the growing threat from amateur and politically motivated hackers, organized crime and other criminal elements, these multi-dimensional threats cannot be ignored.

Based on a thorough understanding of the threats a company faces, executives can make better decisions on how to manage risks in terms of their own business requirements.

In the real world of online business, not all threats can be prevented, nor does it make good business sense to attempt to do so. Therefore organizations need to be prepared to deal with the inevitable unforeseen incident as it occurs. The first step in managing an incident is to quickly detect that it is occurring. Only then can damage be rapidly mitigated. To effectively manage electronic incidents, companies should:

- analyze system organization and network topology
- identify weaknesses in physical, theoretical, hardware and software security, including firewall applications and conventions and virus controls
- review security policy documents and covertly test the policies at workstation level
- develop and implement the necessary mechanisms and controls to identify unusual or unauthorized activity
- develop effective, secure and accurate communication plans, action plans and hierarchies of decision-making in the event of a serious problem; roles and responsibilities should be clearly defined

- develop action plans for scenarios and evaluate the cost and impact of reactions against the impact of specific threats; ill-prepared actions often cause as much damage as the original incident. Action plans should address when and how to involve or respond to outside influences such as press enquires, legal actions, law enforcement and communication or extortion attempts from the perpetrators.
- communicate with and train staff on an ongoing basis regarding what they should do in the event of an incident and to whom they should report suspicious activity; and
- assess incident response plans on a regular basis to ensure they are effective; threats, technology and business are constantly changing

Investigating vs. mitigating

More often than not, when an incident occurs, there are many unanswered questions:

- Who is doing this and why?
- How did he or she get into the systems?
- Is it an employee or contractor (i.e. insider) or an outsider?
- Does he or she have other means of gaining unauthorized access or causing damage?
- What will the perpetrators' reaction be to attempts to identify them or lock them out?
- Has anything been stolen, deleted or otherwise compromised?

Effective decision-making during an incident often requires answers to these basic questions. The answers may be difficult to obtain and require significant investigation. Many company executives find themselves in a dilemma between investigating the incident to answer these questions and simply attempting to stop activity as quickly as possible.

Investigating computer crime is a complicated and highly skilled activity. It requires extensive technical training and expertise in such areas as data forensics and electronic evidence collection and handling. It often requires the ability to effectively interview potential suspects.

Additionally, investigating an incident often requires allowing potentially damaging actions to continue while evidence is collected and activity is traced. If third parties are involved, there may be questions of liability.

These are just a few of the complicated issues that companies need to address when managing a threat. Management should not be thinking about what to do for the first time as it faces its first real incident. If an organization's leaders have not proactively planned how they would manage a problem, it often costs twice as much or more to investigate and resolve the incident, with less than half a chance of successfully determining who is responsible.

Forethought and planning can make a significant difference in the outcome of any crisis.

Control RiskMap (2002) [Online] Available:
http://www.crg.com/press/release11.htm

References and additional readings

Barton, L (1993) *Crisis in Organizations: Managing and Communicating in the Heat of Chaos*, Cincinnati, OH: Southwestern Publishing Company.

Benedict, AC (1994) 'After a Crisis: Restoring Community Relations' in *Communication World*, September.

Benoit, W (1994) *Accounts, Excuses and Apologies: A Theory of Image Restoration Discourse*. Albany: State University of New York Press.

Bergman, E (1994) 'Crisis? What Crisis?' in *Communication World*, April.

Birch, J (1994) 'New Factors in Crisis Planning and response' in *Public Relations Quarterly*, Spring.

'Community Relations: A Necessary Ingredient in Cleanups' in *Environmental Managers*, April, pp. 3–7.

Corzine, R (1996) 'Shell Faces Up to Distrust in Delta Province' in *Financial Times*, July 6.

Corzine, R (1996) 'Shell Plays for High Stakes in Nigeria' in *Financial Times*, July 8.

Dyer, S (1993) 'The Story of a Community Relations Fiasco' in *Public Relations Quarterly*, Summer, vol.38, pp. 33–35.

Fearn-Banks, K (1996) *Crisis Communication: A Casebook Approach*, New Jersey: Lawrence Erlbaum.

Fink, S (1986) *Crisis Management: Planning for the Inevitable*, New York: AMACOM.

Gottschalk, J (ed.) (1993) *Crisis Response: Inside Stories on Managing Image Under Siege*, Detroit, MI: Visible Ink Press, Gale Research.

Green, PS (1994) *Reputation is Everything*, Burr Ridge, IL: Irwin Press.

Jackson, D (1993) 'Bayer Mobilizes Resources to Counter Crisis at Home' in *Chemical Weekly*, April 21, vol. 152, pp. 24–31.

Johnson, DG (1993) 'Crisis management: Forewarned is Forearmed' in *Journal of Business Strategy*, March/April, vol.14, pp. 58–64.

Lerbinger, O (1986) *Managing Corporate Crisis: Strategies for Executives*, Boston: Barrington.

Mortished, C (1996) 'No Longer in Glorious Isolation,' in *The London Times*, July 8.

Peterson, B (1993) 'Crisis Impact on Reputation Management' in *Public Relations Journal*, November, vol.49, p. 28.

Pinsdorf, MK (1987) *Communicating When Your Company is Under Siege: Surviving Public Crisis*, Lexington, MA: Lexington Books.

Reish, MS (1994) 'Chemical Industry Tries to Improve Its Community Relations' in *Chemical and Engineering News*, February.

Shell, A (1993) 'In a Crisis, What You Say Isn't Always What the Public Hears' in *Public Relations Journal*, September, vol.49.

Wilson, J (1985) 'Managing Communication in Crisis: An Expert's View' in *Communication World*, December.

Selected newspaper and magazine articles on the Shell crisis from September 1995 to April 1997:

Financial Times

The Daily Telegraph

The New York Times

The Wall Street Journal

The London Times

The Economist, 2 December 1995, vol. 337, no. 7943, p.18.

The New Statesman & Society, 17 November 1995, pp. 14–15.

Oil and Gas Journal, 31 March 1997, vol. 95, no.13, p. 31.

22 April 1996, vol. 94, no.17, pp. 32, 34, 42.

13 May 1996, vol.94, no. 20, p. 42.

13 February 1995, vol. 93, no.7, p. 3.

29 May 1995, vol. 93, no. 22, p. 20.

Village Voice, 21 November 1995.

Audubon, March–April 1996.

Internet websites

http://www.shellnigeria.com/

http://www.greenpeace.org/

http://www.sierraclub.org/

http://www.cnn.com/

http://sbweb2.med.iacnet.com/infotrac/

http://london.nigeria.com/

Index